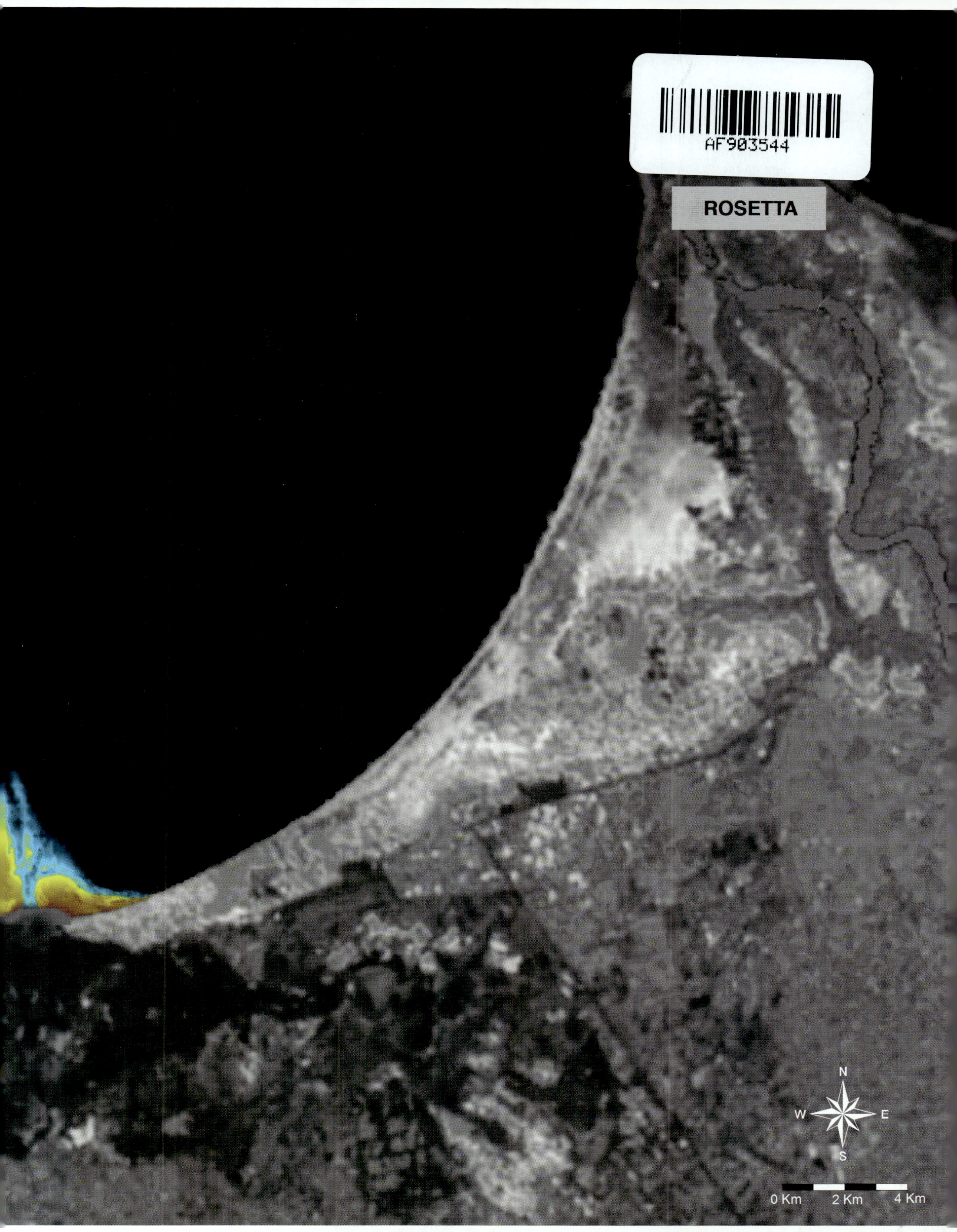
ROSETTA
N
W
E
S
0 Km
2 Km
4 Km

Ship 17: a *baris* from Thonis-Heracleion

Oxford Centre for Maritime Archaeology Monograph Series

Series editors: Franck Goddio and Damian Robinson

1 Topography and Excavation of Heracleion-Thonis and East Canopus (1996–2006)
by Franck Goddio

2 Geoarchaeology
by Jean-Daniel Stanley et al

3 The Naos of the Decades
by Anne-Sophie von Bomhard

4 La stèle de Ptolémée VIII Évergète II à Héracléion
by Christophe Thiers

5 Alexandria and the North-Western Delta
edited by Damian Robinson and Andrew Wilson

6 Maritime Archaeology and Ancient Trade in the Mediterranean
edited by Damian Robinson and Andrew Wilson

7 The Decree of Saïs
by Anne-Sophie von Bomhard

8 Thonis-Heracleion in Context
edited by Damian Robinson and Franck Goddio

9 Theological Defences at the Canopic Gate in the Saïte Period
by Anne-Sophie von Bomhard

10 Ship 17: a *baris* from Thonis-Heracleion.
by Alexander Belov

Ships and boats of the Canopic Region in Egypt

Ship 17
a *baris* from Thonis-Heracleion

by
Alexander Belov

with a contribution by Catherine Grataloup

Oxford Centre for Maritime Archaeology: Monograph 10
School of Archaeology, University of Oxford
2019

Published by the
Oxford Centre for Maritime Archaeology
School of Archaeology
36 Beaumont Street, Oxford OX1 2PG

ISBN 978-1-905905-36-2

British Library Cataloguing in Publication Data.
A catalogue record for this book is available from the British Library.

HILTI The Foundation.

Front cover: an artistic imagining of the complete plan of ship 17 by Christoph Gerigk. The upper half of the model, rendered as a photomosaic, illustrates ship 17 as excavated. Below this, unexcavated areas are mirrored from excavated areas to produce a complete outline for the vessel using the digital elevation model of the ship.

Back cover: Clockwise from the top left: pole no. 12 viewed from the outside of the hull; tenon that passed through the tenon and garboard; lead container discovered near keel segment K1 at the stern; view of the inner hull at the starboard (all photos: C. Gerigk; (c) F. Goddio/IEASM).

Designed and produced by Oxford Book Projects
Printed in Great Britain by Short Run Press, Exeter

For my mother with love

Contents

List of figures

List of tables

Acknowledgements

This book appears as a result of a row of incredible and, I could say, random events, if it were not my belief that there is nothing random in this world.

A considerable part of my scientific career was devoted to physical oceanography. It was my love of the sea and ships that brought me on board of numerous research vessels that ploughed the seas of many names and many colors starting with the monochrome ones—the Black and White. When I came to the realisation that the scientific focus of my field stood between me and nature, it was the sea again that came to my rescue. I started taking interest in sea exploration and shipbuilding history and joined the *Centre for Egyptological Studies of the Russian Academy of Sciences* created by my mother, Dr Galina Belova, in 1999.

If I have achieved anything in science I owe it to my mother who was always there to help and to support me in my endeavours, but never imposed. Her scientific erudition and intuition combined with great love and devotion for ancient Egypt are impressive and they were always good examples for me. It was by her recommendation that I was accepted for the first season of excavations on the sites of Canopus and Thonis-Heracleion that were started by Franck Goddio and his *Institut Européen d'Archéologie sous-marine (IEASM)* in 1999.

It was another great step in my maritime adventures that led me under the surface of the seas that I was so interested in. Being a highly inexperienced diver with hardly a dozen dives logged in the pool, I suddenly found myself in the open-sea and part of a team of real sea wolves, high-class professionals, knowing the laws of the sea and that it never forgives mistakes. My sincere gratitude goes to Franck Goddio, who initiated me into underwater archaeology, and to the members of his team that became and always remains mine. I can only admire the skill and intuition of Jean-Claude Roubaud, the unrivalled master of underwater reconnaissance, who suggested many interesting ideas during the interpretation of the vestiges of the hull of ship 17 from Thonis-Heracleion. Eric Smith and Stéphane Brousse were always very helpful during my first hesitating underwater steps. I address my warmest words of gratitude to my friends Gregory Dalex, Gildas Lesouef, Mario Rodriguez Barrera, Olivier Berger, Yoann Jean, Arnaud Roy, Philippe Rousseau, Patrice Sandrin, Christoph Gerigk, Roland Savoye, Pascal Morisset and Daniel Visnikar. During my first missions in Egypt I was working with Fernando Pereira† and Alain Peton and I keep the most pleasant recollections of those days. It is always a pleasure to take part in underwater excavations when the logistics are provided by Bernard Camier. I would like to thank Sophie Lalbat and Tatiana Curchod for their constant attention and readiness to help in a multitude of organizational questions.

Franck Goddio gave me the great privilege of participating in the excavation of the first ancient Egyptian ship underwater. These excavations turned out to be a fascinating adventure and everybody on board, Franck in the first place, followed them with a vivid interest. The record of the excavation was done by Patrice Sandrin, the draftsman of IEASM, who produced a number of high-precision plans and sections and also offered many interesting hypotheses concerning naval architecture. Christoph Gerigk confirmed his talent once again documenting all stages of excavations with high-quality photographs and also in obtaining photomosaic coverage of the wreck in low-visibility conditions. With the radiocarbon dates being so wide, the analysis of ceramics by Catherine Grataloup became crucial for dating of the ship. I am obliged to her for a chapter on the subject that is incorporated into this book.

I would like to express my sincere gratitude to all of the participants of IEASM missions and to the crew of RV *Princess Duda* (Valetta) who constantly assured our security and comfort on board. The mention of this last name awakes deep emotions in me so strong are the bonds that link all members of Thonis-Heracleion's odyssey to this ship, so familiar and homely.

The next stage of my work brought me to the University of Bordeaux Montaigne in France. I owe my deep and sincere gratitude to Dr Robert Vergnieux, former director of *Archéovision (UPS 3551)*, who kindly agreed to supervise my thesis devoted to the construction of ship 17. He was unwavering in his support during the five years of my study, constantly providing me with his original ideas based on his deep understanding of the principles of Egyptian architecture and art. Robert is the best example of an open-minded and penetrating scholar who is always ready to share his vast knowledge with the others. I am especially grateful to him for giving me access to unique and fascinating iconographical material dating to the end of the eighteenth dynasty from the temple of Aton in Thebes.

Several people among the personnel of *Archéovision* also deserve special recognition in connection with my first approach to the problem of a 3D reconstruction of the hull of ship 17. I would like to express my gratitude to Caroline Delevoie, Pascal Mora and Claire Estoup for their unfailing help and hospitality.

It is with great pleasure that I recollect my two visits to the *Centre Camille Jullian* at the University of Aix-Marseille and the enormous help of Emeritus Professor Patrice Pomey, who readily shared with me his immense experience in ancient shipbuilding and his articles on the subject of Egyptian naval architecture. He kindly corrected my numerous errors and misinterpretations of the construction of ship 17 and offered many interesting suggestions and ideas. His participation in my thesis committee allowed me to considerably improve both the manuscript and this book. Pierre Poveda of *Centre Camille Jullian* patiently clarified with me the basic principles of 3D reconstructions.

I am very happy that working on my thesis allowed me to make the acquaintance with Professor Pascal Arnaud from the University of Lyon Lumière who immediately gave me access to his unpublished article devoted to the ship types that existed on the Nile in antiquity. His vast knowledge of ancient navigation and experience of sailing were always there to help me better understand the navigation space of the ship I studied. I also express my sincere gratitude to him for encouraging me to propose a new translation of the fragment of Herodotus that proved to be very important for the current study.

I had the rare opportunity of having Professor Christophe Thiers from the University of Montpellier as the Chairman of my thesis committee. His valuable critiques touched upon many Egyptological aspects of the manuscript that I corrected with his help. Any remaining errors are my own.

I would like to express my gratitude towards the team members of the Oxford Centre for Maritime Archaeology (OCMA) of Oxford University who joined the excavations on the site of Thonis-Heracleion. My special thanks are for Dr Damian Robinson, the director of the Centre, who proposed interesting ideas concerning the construction of Thonis-Heracleion's ships and who was of great help for the publication of this book, and to Carlos Cabrera Tejedor who guided me in performing a preliminary identification of the wood samples from ship 17. I am obliged to Dr Linda Hulin, who was very helpful with the editing of the English version of the manuscript.

I address my gratitude to Juan-Pablo Olaberria who kindly discussed with me his views on some aspects of hull's stiffness.

The thesis, written in French, was edited by Bernard Boismoreau, Sabine Laemmel and Florence Corrado-Kazanski to whom I express my sincere appreciation.

I have worked in the Centre for Egyptological Studies of the Russian Academy of Sciences for twenty years and take part in the Centre's annual archaeological missions in Memphis and Alexandria. I am grateful to all mambers of the Centre for their support and valuable advice, particularly to the director, Sergei Ivanov.

Last but not least, I would like to offer my thanks to my wife Maria Belova and to my sons Sergei and Dimitry for their patience and support.

1 Egyptian shipbuilding during the Late Period

Introduction

The three centuries of the Late Period (664–332 BC) compromise a comparatively short period in the more than four thousand-year history of naval architecture in ancient Egypt. It is also a period during which, prior to the discovery of Thonis-Heracleion and its extensive nautical assemblage, there was a dearth of evidence. It is consequently the purpose of this chapter to set out the material previously used to characterise vessels in the Late Period and especially the Egyptian *baris*-boat. This will provide the essential background to the subsequent discussions of ship 17 and its significance.

Archaeological evidence for Egyptian shipbuilding in the Late Period

Before excavations of Thonis-Heracleion began only one vessel dating to the Late Period had been found. The Mataria boat was discovered in 1987 during construction work near Heliopolis, now a suburb of Cairo.[1] It has been dated to 2450 ±50 BP (*c.* 500 BC), which is around the time of Herodotus' visit to Egypt, when he saw a *baris*-boat being built (below). Unfortunately, the Mataria boat was excavated rapidly under salvage conditions and the documentation of it was not ideal. For example, the plan—or rather, the sketch—of the planking, has obvious graphical distortions. Despite these shortcomings, it provides an idea of the arrangement of the planking, of the shape and dimensions of separate planks, and on the wood species used in the construction, which was mainly local sycomore fig (*Ficus sycomorus*) wood.[2]

The preserved part of the hull was 8 m long and 4 m wide, although initially the boat must have been more than 11 m long.[3] The length of the planks varied from 0.8–6 m while the majority was between 2 and 4 m in length. The planks, including the keel plank, were about 20 cm wide and 11 cm thick. The planking was assembled by tenons that exceeded 17 cm in length, although none of them were preserved completely. Some of the tenons were secured to the planking by tenon pegs 2 cm in diameter. According to Ward, the Mataria boat 'reflects Egyptian nautical traditions that incorporated a feature of Mediterranean shipbuilding'.[4] Ward concludes:

> *The large tenons of Mataria continue an ancient tradition; the pegs might reflect shipbuilding techniques adopted by the Egyptians in the late New Kingdom (c. 1250 BC). We must distinguish, however, between the Mediterranean practice of locking every joint in place with pegs on either side of a plank seam and the Egyptian practice of using pegs, or treenails in some joints.*[5]

Iconographic evidence for Egyptian shipbuilding of the Late Period

While there are many representations and models of ships and boats dating to the Old and Middle Kingdoms, they become rare during the New Kingdom and sporadic in the Late Period. As Landström noted, 'pictures of ships from this period are few, coarsely executed, and—in comparison with what had gone before—fairly uninteresting'.[6]

The exception to this are five sandstone wall fragments known as 'the blocks of Piankhy', currently in the Egyptian Museum in Cairo,[7] which date to the twenty-sixth dynasty (Psamtik I, 664–610 BC).[8] The reliefs on the blocks represent the arrival of the fleet in Karnak and two of the ships can be classified as large river-faring vessels of *ḏ3y*-type.[9] According to the accompanying inscription they are 'the ship of Amun' and 'the great ship of Saïs' (Figure 1.1).

Figure 1.1 Representation of two *ḏ3y* boats on the 'blocks of Piankhi' (block 1): 'the ship of Amun' (right) and the great ship of Saïs' (left). (Photo: Brugsch Bey. From Benson and Gourley 1899: 320, pl. 20.)

1 Haldane 1993: 240–50; 1996a: 235–44; 1996b: 441; Vinson 1994: 47–8; Ward 2000: 129–35.
2 Ward 2000: 129.
3 Ibid.: 131.
4 Ibid.: 133.
5 Ibid.: 134.
6 Landström 1970: 140.
7 JE 31886.
8 Perdu 2004: 225–40; 2012: 151–7.
9 Jones 1988: 151.

Figure 1.2 *wsḫ*-ship on the third 'Piankhy' block. (Photo: A. Lecler; © IFAO.)

Both ships have a crescent-shaped hull but they are not exactly the same. While the stern of 'the great ship of Saïs' is greatly elevated, this is not the case with 'the ship of Amun' because of the decorative quasi-vertical timber at the stern. The first boat is propelled by 20 rowers (ten at the bow and ten aft), the second by 12. 'The great ship of Saïs' seems to be of more imposing dimensions, a conclusion that is supported by its greater length in relation to the other vessel, by the number of rowers, and by the presence of a large superstructure in the middle of the ship. In addition, this vessel demonstrates two levels of through-beams, thus attesting to a considerable depth of the hull. Both ships are steered by large axial rudders.

The text on the third block indicates that the third ship depicted on it belongs to a type of a *wsḫ*-ship (Figure 1.2).[10] The term *wsḫ* is paronymous to the word 'width' in ancient Egyptian, thus likely indicating a barge.[11] Such a vessel would have had a breadth to length ratio of about 1:3 that would provide considerable freight-carrying capacity.[12] The third ship has a slightly different form of hull. Even if the stern is rather high, the curvature of the hull is less pronounced in comparison to 'the great ship of Saïs'. This boat has no deck superstructures and is steered by an axial rudder. The traces on blocks two and four suggest that the relief initially contained a dozen *wsḫ*-ships.[13] Their crew consisted of about 20 rowers, a helmsman and an ensign-bearer.

The shortage of iconographic evidence firmly dated to the Late Period resulted in the widening of the search for possible parallels to both earlier and later periods of Egyptian history. It seems that during the Ptolemaic Period (332–30 BC) ships were built not only according to Mediterranean standards but also continued to be built along traditional Egyptian lines, especially on the Nile.[14] The craftsmen of the Ptolemaic Period continued to uphold the artistic standards of pharaonic Egypt and the appearance of the ritual boats represented in the temples continued to be exactly the same as many centuries before, making ships built in a Mediterranean-style difficult to trace. Some indications of this, however, may

10 'Boat for transporting goods, cargo-boat, barge'. See Jones 1988: 135.
11 Meeks 1994: 257.
12 Perdu 2004: 226.
13 Ibid.
14 Casson 1995: 340; Pomey 2015b; Vinson 1994: 49.

be seen in a model from Tebtunis,[15] which has a lightly rockered keel and a hull shape that resembles that of the Kinneret boat (*c.* 100 BC–67 AD) from the Sea of Galilee in Israel. This vessel was built between the 1st century BC and 2nd century AD.[16] The cedar planks were edge-joined to each other by means of pegged mortise-and-tenon joints and fixed to the frames by iron nails. The framing consisted of regularly alternating floors, futtocks and half-frames.

The Nile mosaic from Palestrina in Italy (ancient city of Praeneste), dated to *c.* 125 BC, has been called 'the catalogue of Nile river-faring craft'.[17] Apart from the representation of a Hellenistic warship, the mosaic contains those of the Egyptian types among which some are identified as the *kubaia*[18] and *baris*.[19] In addition, there are also images of small papyrus boats and a woven boat (*paktôn*).[20] It has been already noticed that the aquatic landscape of the mosaic corresponds well to the descriptions of the Delta region and at the same time to the topography of Thonis-Heracleion.[21]

Literary evidence

In the ancient literary evidence two main discussions come out of that from the Late Period, which concern both military and cargo vessels. While those pertaining to the latter are obviously the most important with regards to ship 17, it is nevertheless instructive to also first consider the modern debate about the types of warships, as this clearly illustrates the limits of literary evidence when dealing with the construction of ancient ships.

In 1970–80s there was an animated debate between Lloyd and Basch on the subject of the types of warships in the fleets of Saïte pharaohs (twenty-sixth dynasty, 664–525 BC). Lloyd suggests that Greek-style triremes were incorporated into the Egyptian navy under Necho II (610–595 BC),[22] and attempted to demonstrate that the term *kbnt*, attested in Egypt since the Old Kingdom, designated a warship during the Persian Period.[23] Lloyd adheres to the epithet of 'fast-running galley' that was first applied to the *kbnt* by Save-Soderbergh[24] and emphasises that the term was definitely applied to triremes during the Ptolemaic Period. The hieroglyph used for the determinative for *kbnt* at the end of the fourth century BC does not have a single feature of an Egyptian boat and the form of the hull and a ram of the ship represented resemble those of the trireme.[25] Lloyd concluded that in the time of Necho II the term *kbnt* applied to a 'Greek-style war galley'. He favours a date for the invention of the trireme of around the middle of the seventh century BC,[26] which would theoretically allow its appearance in the Necho's fleet through their Corinthian allies. Lloyd's conclusions,[27] however, contradict those of Basch,[28] who claims that the triremes of Necho were of Phoenician origin,[29] basing his argument upon some construction features of the Phoenician trireme, such as the absence of the outriggers (*parexeiresia*) and its larger hull in comparison to the Greek trireme.[30] He underlines the close and long-standing relationship between Egypt and Phoenicia and the presence of Phoenician shipbuilders in Peru-nefer at Memphis, the principal Egyptian shipyard from the New Kingdom. Lloyd, in turn, was not convinced that Phoenician shipbuilders were still present in Peru-nefer in the time of Necho II. He equally denies any pronounced constructional differences between Greek and Phoenician triremes.[31] Closing the discussion, Basch analyses the numismatic material that seems to confirm the absence of outriggers on Phoenician triremes before Alexander the Great, but concludes that the Phoenician origin of the trireme in Egypt remains a hypothesis.[32]

More recently, Darnell has returned to this subject and after considering available Egyptian and Greek written sources he concludes that *kbn.t* describes seaworthy vessels, be they vessels of war, trade or exploration. He also concludes that there was no Egyptian word denoting 'trireme' specifically, or 'warship of foreign design' in general, even during the Ptolemaic Period, when Egypt's navy was amongst the largest in the Hellenistic world. Darnell consequently suggests that any equation of the Late Period *kbn.t* and the trireme/warship is incorrect.[33]

What this debate most clearly illustrates is the acute absence of archaeological evidence beyond relatively indistinct images of warships on coins. In the absence of a wrecked or abandoned Egyptian trireme from which to study the crucial details of the nautical architecture, the uncertainty surrounding their precise form or forms in the Late Period continues. Even the representation of the galley on the Nile mosaic from Palestrina discussed above is not particularly helpful, as warship design had moved on apace and what it represented on the mosaic is a later type of bireme.[34]

15 Phoebe Apperson Hearst Museum of Anthropology, CA, USA, PAHMA 6-21372; see Merriman 2011: app. C.
16 Steffy 1990: 41; Wachsmann et al. 1987: 243.
17 Pomey 2015b.
18 Basch 2008: 74; Pomey 2015b: 161–4.
19 Pomey 2015b: 164–6.
20 Ibid.: 156–60.
21 Goddio 2007: 128–30.
22 Lloyd 1972a; 1972b. This hypothesis was subsequently supported by Van't Dack 1978 and Jones 1988: 148.
23 Lloyd 1972a: 272.
24 Save-Soderbergh 1946: 48 note 5.
25 Lloyd 1972a: 275.
26 657–583 BC according to Thucydides *De bello peloponnesiaco* 1.13.
27 Also see his later article: Lloyd 1975a.
28 Basch 1969: 142.
29 This point of view is supported by Casson 1970 and 1971: 115.
30 Basch 1977.
31 Lloyd 1980.
32 Basch 1980.
33 Darnell 1992: 89.
34 Basch 1987: 441–2; Pomey 2015: 167–8.

Table 1.1 Greek papyri containing information on the *baris*.

Papyrus	Title	Provenance	Date	Mention of a ***baris***	Comments
P.Hib. 1 100 V	Account: a receipt for rent	Herakleopolites	267 BC	ọ vac.? σιτομέτρης Ξάνθου παρ' εὐφρά[ν] ορος δι' Ἀντιπάτρου εἰς τὰ ἐκφόρια τοῦ Ἀλεξάνδρου κλήρου εἰς τὸ ιθ (ἔτος) ἐξ Ἀνατ̣ιεὺ κριθῶ̣(ν(?)) (ἀρτάβας) λ̣ π̣ρ̣( ) εἰς **βᾶριν** ἐφ' ἧς κυβερνήτης vac.? ναύκληρος vac.?	Lines 10–15: *baris* loaded with barley
P.Cair.Zen. 4 59745	Account: payments to workmen	Philadelphia? (Arsinoites)	255–254 BC	δι' ἐγγύου Πάσιτος κυ(βερνήτου) τῆς λιθηγοῦ **βάρεως**	Line 66: λιθηγὸς βᾶρις : a stone-hauling *baris*
P.Coll.Youtie 1 7	Letter from Amadokos to Klion	Magdola or Ghoran (Arsinoites)	224 BC	κυβερνήτης Ἐροβάστις, ἀναπλέον κατὰ Τμοιενέτιν περικόψαι **βᾶριν** ἐν ἧι ἐπέπλεον γυναῖκες	Lines 4–6: a *baris* transporting women involved in a collision
W.Chr. 11 A	War between Hermonthis and Krokodilopolis	Krokodilopolis (Pathyrites)	123 BC	σὺν [ὅπλοις κ]αὶ ἐπαγαγόν[τες κατ' αὐ-] τὴν τὴν πόλι[ν . . . . ἐν] **β̣άρει**	Lines 21–22: *baris* transporting soldiers
P.Tebt. 3 .1 701	Register of official business	Tebtynis (Arsinoites)	235 BC	\[Ἀρχι]τίμωι./ [σύντα]ξον ἐμβαλέσθαι εἰς τὸν Ἀ̣θ̣ὺ̣[ρ ἐ]ν τ̣ῆ̣ **βάρ̣ι̣**	Line 26: cargo loaded in a *baris*

The most important source on Egyptian shipbuilding in the Late Period is undoubtedly Herodotus. Written around 450 BC, one paragraph of his *Historia* (2.96) describes in detail the construction of a type of Egyptian river freighter called a *baris*. The importance of these twenty-three lines, which contain the description of the construction sequence of the *baris* and mention the wood species used, is far greater than the scarce iconographic material from the Late Period and will be discussed in detail in Chapter 10. What is significant here is that like the discussion of the warships, interpreting exactly what Herodotus was describing was difficult prior to the excavation of ship 17 and then ship 43 from Thonis-Heracleion.

The Greek term *baris* (βᾶρις) probably originates in the ancient Egyptian boat type called *br* (*byr*),[35] a term which itself may have had a foreign origin.[36] When it first appears in the eighteenth dynasty, *br* refers to a sea-going craft.[37] This type of boat is frequently mentioned in Egyptian sources,[38] indicating that it was a rather common word for a transport ship at this time.[39] Demotic texts mentioning *br*-ships, however, are not numerous and contrary to the hieroglyphic texts most of them probably refer to Nile cargo vessels.[40] Vinson also cites two documents that can indicate a military use of *br*-ships,[41] with Darnell suggesting a parallel between the *br*-ships of the defensive fleet of Ptolemy V mentioned on the Rosetta Stone (196 BC) and those employed by Ramesses III (1187–1157 BC) to defend the Delta against the Sea Peoples.[42] Another interesting document is *P.Cowley* 26 from the Persian Period (412 BC) that contains an inventory for a *br*-boat built of cedar.[43]

In Greek literature, the *baris*[44] was connected with Egypt from the time of Aeschylus (525–456 BC).[45] Textual evidence from Greek papyri suggests that it was primarily a freighter (Table 1.1.),[46] with Casson noting the all-purpose nature of the craft, with cargoes of fish, or stone, as well as their use carrying soldiers documented in the literary record.[47]

Herodotus' description of the *baris* is part of the general section concerning various aspects of life in the Delta.[48] According to Vinson it is likely that this type of vessel developed in the Delta, rather than in the Nile valley:

> *It is probably not without interest that II, 96 is the fifth of a series of chapters which deal with aquatic life in Egypt, a train of thought which begins as Herodotus winds up a section on the customs of the Egyptians who live in the Nile Valley. Beginning with Chapter 92, Herodotus turns to the marshes of Lower Egypt, a topic which leads him to talk about food in the marshes (Chapter 92), fish which live in Delta lakes (Chapter 93), the types of oil which are available to marsh dwellers (Chapter 94) and the abundance of insects in the Delta (Chapter 95). Thus, it cannot necessarily*

35 Casson 1971: 341 note 64; Vinson 1998: 251–60.

36 Vinson 1993: 147 note 69 and 70; 1994: 44–5.

37 Casson 1971: 341 note 64; Vinson 1994: 44–5; 1998: 252.

38 Jones 1988: 136.

39 Vinson 1994: 45.

40 Ibid. 1998: 252–3. The author cites the following ostraca: *O.Zürich* 1867 (Roman); *O.Leiden* 136, 2; *O.Uppsala* 1185, 2; *O.Leiden* 340, 3.

41 Ibid.: 253. Line 12 of the Rosetta Stone contains *νῆες* (ships) in the Greek text that correspond to the hieroglyphic *kbn.wt* and to the Demotic *byry*. Another example cited is the Roman *P.Krall* 14/8 mentioning *br*-ships as part of a naval fleet.

42 Darnell 1992: 72–3 notes 21 and 54.

43 Ray 1988: 268; Vinson 1998: 253. It is not only the imported cedar wood used in the construction of this boat that attracts one's attention, but also the great quantity of nails. Bronze nails were employed in the construction of some of the larger ships of Thonis-Heracleion for attaching the frames to the planking. This is the case of ship 61 (234±40 BC) and of ship 11 (400–325 BC). These boats demonstrate a well-developed internal structure and seem to belong to a Greco-Roman type of construction.

44 βᾶρις, ιδος (also-εως), Ion. ιος, ἡ: flat-bottomed boat, used in Egypt. See Liddell 1940. According to the Greek-Russian Lexicon edited by A. D. Veisman (Вейсман 1899: 243), βᾶρις is a term that specifically refers to an Egyptian boat or ship or, in a more general sense, to a ship of a non-Greek tradition. Cf. 'βαρβάρους βάριδας' (Euripides *Iphigenia in Aulis* 297) and 'βάρβαροι βάριδες' (Procopius *De aedificiis* 1.6).

45 Aeschylus *Supplices* 873; Arnaud 2015: 115.

46 Casson 1971: 340–41, notes 60–4; Vinson 1998: 254.

47 Ibid. 1971: 340 note 64.

48 Herodotus *Historia* 2.60–63, 92–95, 97.

be taken for granted that Herodotus is discussing ship construction practices which were current throughout the Nile Valley. Perhaps Herodotus was discussing boats he had seen in the Delta. The term βᾶρις-boat, which is the name Herodotus gives to the boats he describes in Chapter 96, is also connected with Delta shipping in 2.179.[49]

Thus, for ship 17 it is likely that we are dealing with a vessel that was locally-made, perhaps in the very shipyard that Herodotus must have visited in order to write such a precise text. While this was vital for the correct interpretation of the remains of ship 17, the nautical architecture of the vessel also helped to clarify many aspects of his description that had previously been opaque. This can be seen, for example, in Lloyd's work translating and providing a commentary of book two of Herodotus' Historia,[50] which also resulted in a later article in which he criticises earlier translations of several lines devoted to the construction of the *baris*.[51] In particular, he insists that ζυγά in the Greek text refers to cross-thwarts, but not to frames. Citing a parallel of a near-contemporary boat of the Nile called the *naggr* (*nuggar*) he favours the translation of the verb ἐμπακτόω as luting and not caulking or lashing proposed in some other translations.[52] However, he denies the possibility of the rudder's stock passing through the keel. Again, like in the discussion of warship, the lack of the actual remains of a *baris* hindered Lloyd's interpretations of the text.

There is a rich corpus of evidence for Egyptian shipbuilding in the Ptolemaic Period.[53] It should, however, be treated with caution as Mediterranean methods of ship construction were extensively used in Egypt at this time. Casson, for example, cites Greek papyri mentioning *ploion Hellenikon* ('Greek boat') and suggests that this term designates a boat built according to Greek tradition, as opposed to indigenous craft that continued plying the Nile.[54] Nevertheless, these documents also demonstrate the persistence of traditional shipbuilding techniques on the Nile. Thus, a Ptolemaic Period papyrus *P.Petrie* 3.46.1 mentions the lashings of the hull to be retightened—an operation which the Mediterranean-type vessels of the time had not required for many centuries.[55] Another papyrus (*P.SI* 4.382) suggests that local acacia wood continued to be used in the construction of vessels.[56]

Ptolemaic papyri contain valuable information on the types of Nile grain carriers in use, on their itineraries, and tonnage.[57] The normal destination of grain ships was Alexandria; however, the 'traders might sail both up and downstream in the search of good markets or better deals'.[58] Small capacity barges called *prosagogides* were mainly employed on the Fayum canals, while the most common freighter of the Nile was the *kerkuros* that was of various tonnage ranging from 1,000 to 18,000 *artabas*.[59]

The *baris* is mentioned several times in Hellenistic documents[60] as a type of vessel alongside some other obviously indigenous types including, for example, a craft made of wicker called a *paktôn*.[61] The latest mention of the *baris* in Egyptian sources comes at the end of the second century BC, after which it disappears, probably replaced by other types of cargo vessels, such as the *kubaia* and *kerkuros*.[62] The latest mention of the *baris* in Classical literature comes from Plutarch in the first century AD, by which time it seems to have lost its initial meaning and is instead used to refer to a boat made of papyrus.[63]

Ethnographic evidence

Clarke[64] and Hornell use ethnographic evidence to suggest that the Nile boat called the *nuggar* or *naggr*, which was still found on the Nile in Sudan in the early twentieth century, retains numerous constructional features of ancient vessels. The *naggr*, a small, multipurpose Nile boat, is traditionally built from the wood of acacia. The flaws of the timber determined the choice of some common construction methods and solutions, even if there certainly exist many differences from the ancient craft. Every stage of boat building is described in detail and as such these ethnographic studies provide a wealth of information with which to reflect upon the evolution of Egyptian riverine architecture.[65]

The potential contribution of ship 17

As the discussion above has emphasised, while there is literary evidence about the *baris* that allows us to understand this type of vessel as a Nile cargo boat and iconographic representations that indicate how it could looked, the lack of any direct nautical remains in the archaeological record effectively placed an upper ceiling on what could be done with this evidence. The discovery of the remains of the hulls of literally tens of ships

49 Vinson 1998: 252.
50 Lloyd 1975b; 1976.
51 Ibid.: 1979.
52 This question will be discussed below in Chapter 10 (luting of the joints).
53 Arnaud 2012; 2015; Casson 1990; Darnell 1992: 68–81, 87; Hauben 1979; Rostovtzeff 1940; Thompson and Garnsey 1983; Vinson 1996: 202; 1997.
54 Casson 1971: 340. He refers to ancient Egyptian naval architecture mainly of the Ptolemaic and Roman Periods in other articles: Casson 1942; 1950; 1990; 2004.
55 See Vinson 1996: 203.
56 248/247 BC, Philadelphia.
57 Meijer 1996; Thompson and Garnsey 1983.
58 Thompson and Garnsey 1983: 71.
59 Ibid.: 72. 'The most commonly recorded size of *kerkuros*, that of 10,000 *artabas*, held 306.5 tons, and the largest–551.7'.
60 See Arnaud 2015: 114–5.
61 Ibid.: 123–4. This type may be followed in the papyri until the eighth century AD.
62 Arnaud 2015a: 111.
63 Plutarch *De Osiris et Isis* 358a; see Arnaud 2015: 115, n. 52; 116.
64 Clarke 1920: 2–50.
65 Hornell 1942: 1–36; 1943a: 27–41; Landström 1970: 140–1.

that were likely to have been *barides* in the waterways of Thonis-Heracleion have swept away this barrier. Instead of relying on literature and iconography bolstered by ethnography, ship 17 offers us the opportunity to place real archaeological evidence into these debates and clarify a number of unanswered questions primarily concerning the construction and a choice of building materials of the ships.

2 The context and date of ship 17

The site of Thonis-Heracleion and its shipwrecks

The port of Thonis-Heracleion was situated on the Mediterranean coast of Egypt about 20 km to the east of Alexandria. The combined actions of several natural phenomena caused the sinking of around 110 km^2 of the ancient Canopic coastline, including the port-city.[1] Thonis-Heracleion is now six and a half km away from the western coast of Aboukir Bay (Figure 2.1) and lies between six and eight metres beneath current sea level.

The city of Thonis-Heracleion is attested by numerous classical sources, starting from the first half of the sixth century BC.[2] Analysis of Herodotus' accounts[3] led Yoyotte to suggest that the *emporion* of Thonis existed from the eighth or early seventh century BC.[4] In the Late Period, the city controlled access to the Canopic branch of the Nile and was engaged in trade with Greece.[5] Numerous archaeological remains and epigraphic finds confirm the identification of the site as the submerged city of Thonis-Heracleion.[6]

Main topographic features of the site

The city as currently defined occupies a rectangular space about 1.2 km^2 in area. The main part of Thonis-Heracleion is located on an island surrounded by a large inland lake to the west and a few basins of smaller size to the east. A chain of basins and a narrow passage between sand dunes connected the city to the Canopic branch of the Nile (Figure 2.2).[7] The Grand Canal runs across the main peninsula just to the north from the temple of Amun-Gereb and links the harbour basins to the western lake.[8] Ports and channels occupy considerable space in Thonis-Heracleion and they contain more than 70 ancient ships[9] and some 700 ancient anchors that testify to the scale of shipping in antiquity.

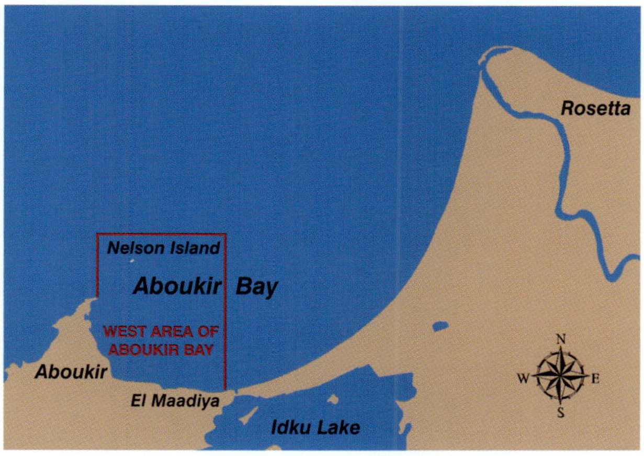

Figure 2.1 The site of the survey and underwater excavations. (Map created by F. Goddio; © F. Goddio/IEASM.)

The distribution of ships in Thonis-Heracleion

The majority of the ships were found in the Central Port and in the Grand Canal (Figure 2.2), where they form several distinct clusters with the hulls touching each other.[10] Five ships were found in the Grand Canal near the remains of quays around the temple of Amun-Gereb (sites H2 and H4). Other vessels were found dispersed around smaller basins and secondary channels in the city, particularly in the area of the south-east port.

The stratigraphy of the site is complicated by geological phenomena such as sea transgression, erosion, landslides, and soil liquefaction.[11] Consequently, the hulls of the majority of the ships are partially embedded in a dense grey clay that provides excellent protection for the wood.

Preliminary dating of 45 ships from the northern part of the port is presented in Figure 2.3. The majority of ships date between the sixth and second centuries BC.[12] About 40 per cent of the ships are probably Ptolemaic (332–30 BC). The remaining ships date to the Late Period (664–332 BC) or, possibly, to the Third Intermediate Period (1069–664 BC).

1 Goddio 2007: 8; Stanley 2007.
2 Cf. Bernand 1970 for a summary.
3 Herodotus *Historia* 2.179.
4 Yoyotte 2001; Fabre 2008.
5 Yoyotte 1958. Summary of major ancient sources concerning Thonis-Heracleion, history of its scientific exploration from the middle of the eighteenth century to the present, and significant results achieved can be found in Goddio 1995; 2007.
6 Yoyotte 2001; Goddio 2007: 69–130; Fabre and Goddio 2008.
7 Goddio 2007: 69–130.
8 Ibid.
9 To date, 70 ships have been discovered in Thonis-Heracleion. Currently this is the largest accumulation of ancient ships in the world. Sixteen ships were found in Pisa (Bruni 2000), 13 in Olba (D'Oriano and Riccardi 2004), and 34 ships in Istanbul (Kocabaş and Kocabaş 2008; Gunsenin 2010; Ingram and Jones 2010; Pulak and Ingram 2013; Kocabaş 2014).
10 Goddio 2007: 69–130.
11 Stanley 2007.
12 Fabre and Belov 2012: 108.

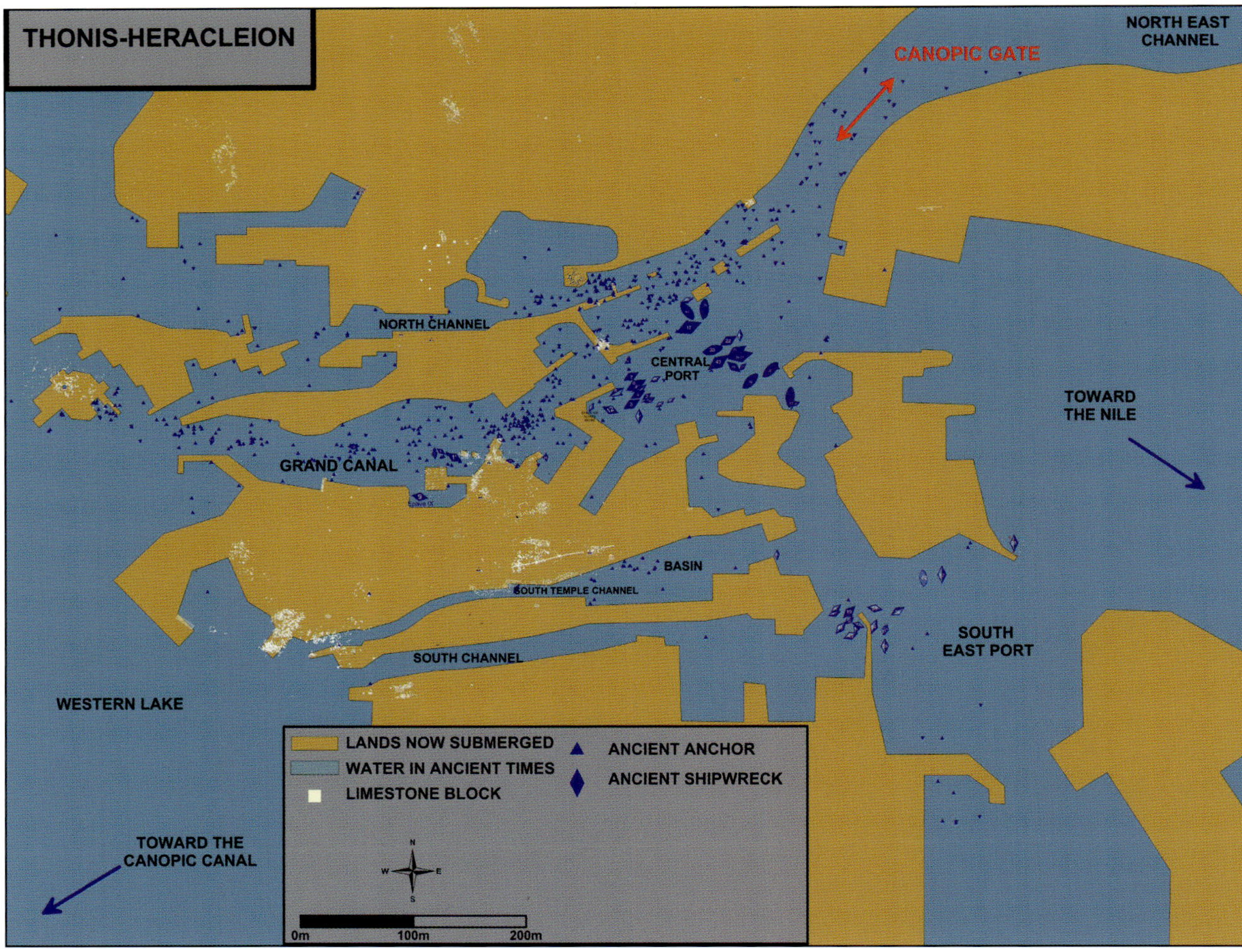

Figure 2.2 Main topographic feature and distribution pattern of ships at Thonis-Heracleion. (Map: F. Goddio; © F. Goddio/IEASM.)

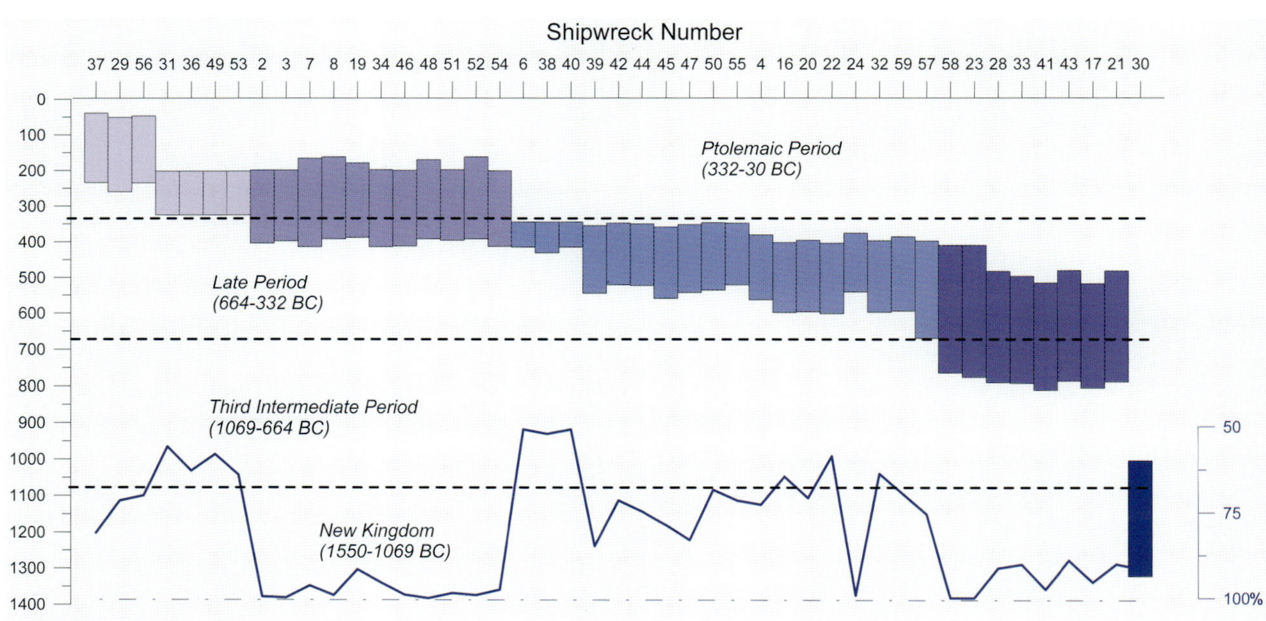

Figure 2.3 Calibrated dates of 45 ships from Thonis-Heracleion and dating probability curve (blue line).

It would appear that the vessels of each cluster are closely dated.[13] The majority of the ships are without cargo and are just empty shells, devoid not only of material culture but of many constructional pieces as well. These were not shipwrecks—catastrophically lost along with all their contents—but ships that were

13 Ibid.: 109; Robinson, Goddio and Fabre 2017.

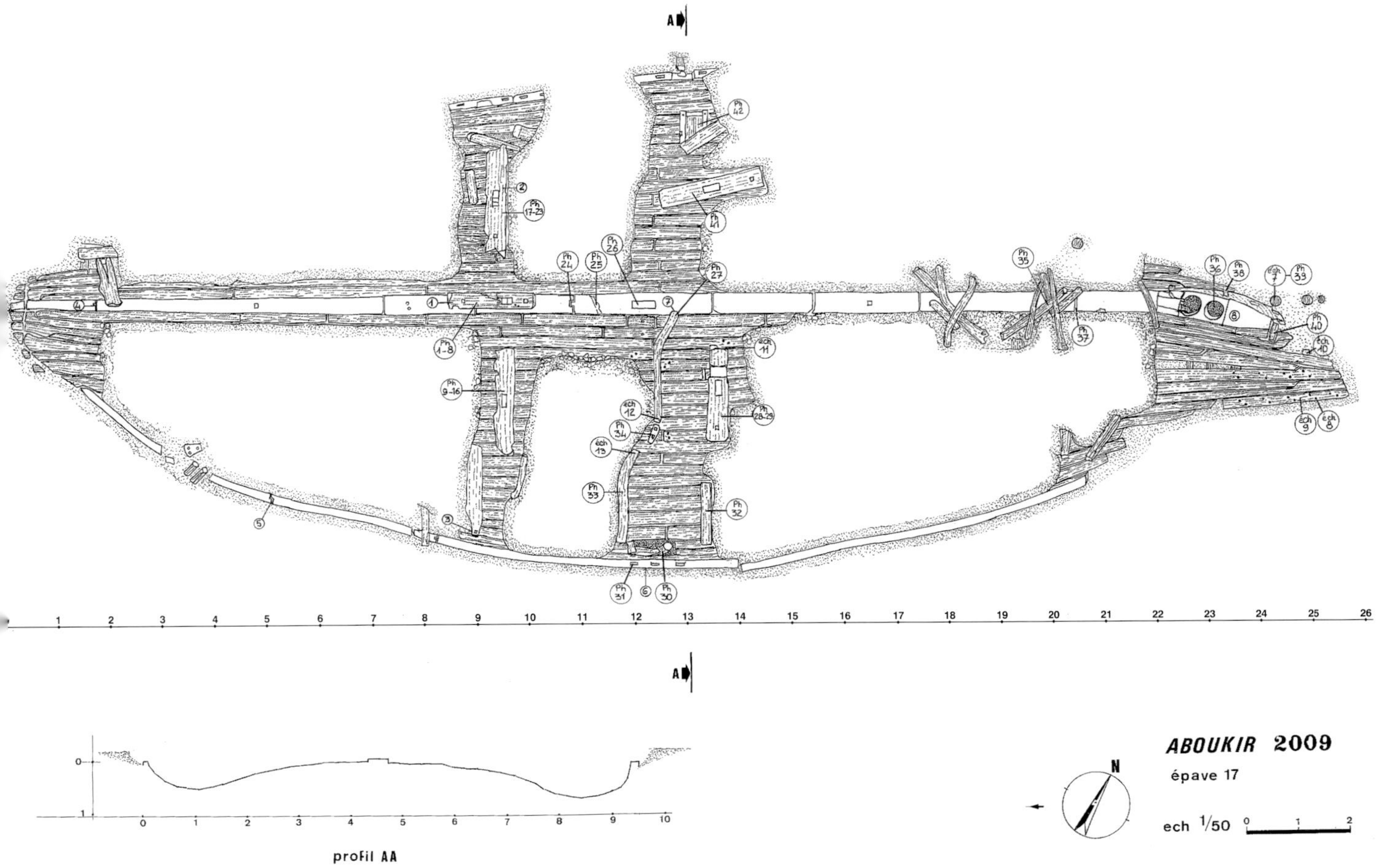

Figure 2.4 General plan of shipwreck 17 in 2009; oriented to the north as the position of the bow had not yet been established. (Drawing: P. Sandrin; © F. Goddio/IEASM.)

deliberately abandoned.[14] Different hypotheses for their origins have been suggested.[15] Many ships are fixed to the bottom with vertical wooden posts and thus the idea of land reclamation or port infrastructure seems very tempting. A more unlikely possibility, particularly for ship 17, is that the ships were scuttled to block access to the port.[16]

Preliminary results suggest that local species of wood (acacia and sycomore fig) predominate over imported species (oak and pine).[17] This would suggest that the ships from Thonis-Heracleion are most likely to have been Egyptian vessels, perhaps even constructed in the shipyard that Herodotus visited during his passage through the port-city.[18]

Ship 17

Ship 17 was discovered during a survey of the main port of Thonis-Heracleion in 2003 and was the first ship from the site to be excavated. Approximately 70 per cent of the surface of the hull was uncovered during three excavation seasons (2009, 2010, and 2011). The shallow depth of the site, averaging 7–8 m, was major hindrance during the excavations as the swell reached the bottom and resulted in low visibility. Visibility of one and a half metres was considered to be normal, while quite often it fell to zero, particularly during the operation of the water dredge. Rare periods of good visibility (5–6 m) were advantageous for drawings and wide-angle photography.

The excavation of ship 17

The 2009 season

The excavation of the ship started on 24th April and continued until 9th May. The upper part of the planking that retained its quasi-vertical position was discovered under approximately 20–30 cm of sand. The line of the planking was followed with the water-dredge and the south-west extremity of the ship uncovered. The position of its central axis was determined by a keel-like element that projected inside the hull. A baseline and a rope grid

14 Robinson 2015. See also Richards 2008 and Richards and Seel 2013.
15 Robinson 2015; 2018.
16 Ancient sources speak of blocking the harbours with a chain of ships; however, they remained afloat: Thucydides *De Bello Peloponnesiaco* 4.12. In modern times there are many examples when ships were scuttled for the same purpose, see Ford 2013: 200–1.
17 Fabre 2011: 16–8; Fabre and Belov 2012: 109–11.
18 Vinson 1998: 252, 260; Herodotus *Historia* 2.96.

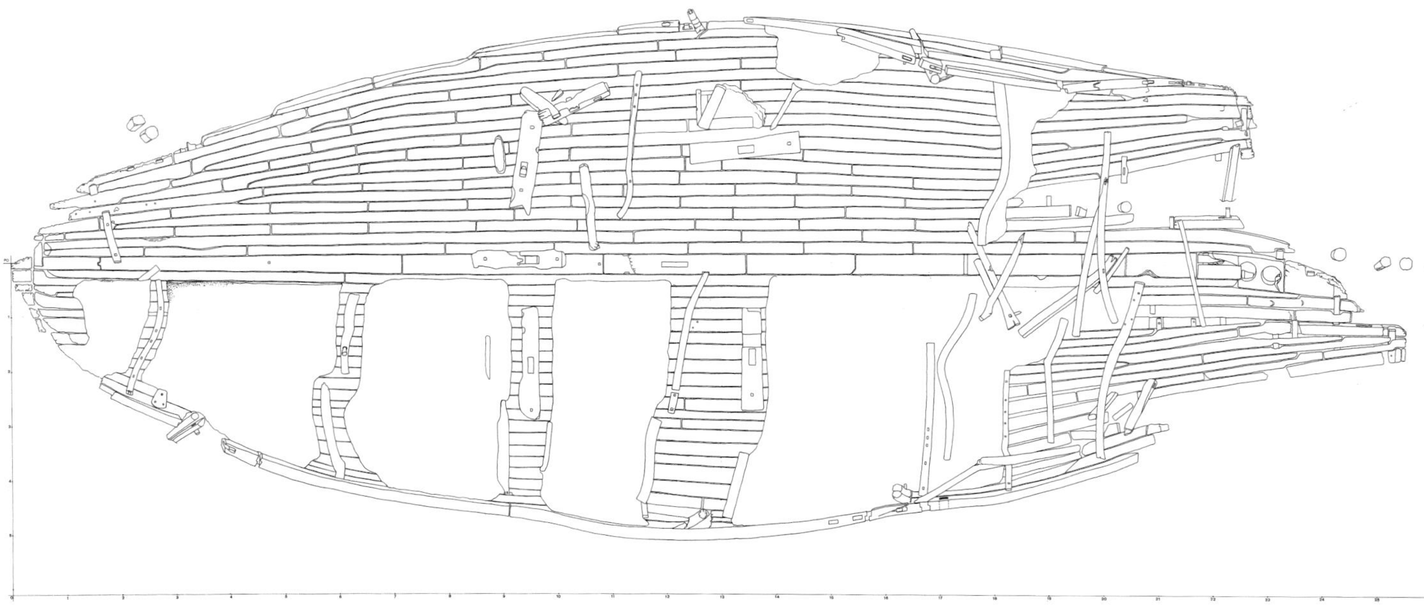

Figure 2.5 General plan of shipwreck 17 in 2010; oriented to the north as the position of the bow had not yet been established. (Drawing P. Sandrin; © F. Goddio/IEASM.)

with one-metre squares were installed over the ship and excavation began following the keel from the south-west to the north-east. A point slightly beyond the preserved end of the bow served as a provisional baseline zero.

After cleaning of the central part of the hull, the position of the stern was firmly established by two openings in the keel that were discovered at the north-east end of the ship. They were used for the passage of the vessel's axial rudder stock and their position indicated that the bow of the ship pointed to 240 compass degrees.

At the end of the season the keel and two neighbouring strakes of each board were completely excavated. In addition, two transversal trenches were placed 9 and 13 m from the bow, revealing numerous details of the internal structure of the ship: supports, bracing timbers, a small hatch, and also details of construction crossing the keel. The southern outline of the preserved part of the hull was established, as well as its maximum preserved breadth amidships (Figure 2.4). Six samples were taken for determination of wood species and radiocarbon dating.[19]

2010 season

The second excavation campaign lasted from 20th April to 15th May, during which the starboard of the ship was completely excavated (Figure 2.5). Towards the end of the season, the stern was cleaned and two new trenches were opened on the port revealing more pieces of its internal structure. Further wood samples were obtained[20] alongside pieces of rope and luting. The first two starboard strakes in the region of the stern were also disassembled to study the joints between the keel segments.

An inventory system was developed for constructional elements that included a code for the type of the piece, followed by its number and an identifier of the board ('S' for the starboard and 'P' for the port) (Table 2.1).

2011 season

There were no large-scale excavations during the mission in 2011, which lasted from 30th September to 9th October, the principal objective of which was to obtain more information on the hull. Thus only two small sections were investigated along the keel and one transversal section on the starboard bow (Figure 2.6).

The central segment of the keel (K6) was temporarily lifted from the water onto RV *Princess Duda*. Water was regularly applied to the wood during the eight hours it was studied, after which the piece was returned to its original position in the ship.

A deep trench was made to follow one of the poles that encircled ship 17 and eight more samples of wood were collected.[21] Favorable weather conditions allowed five

Table 2.1 Inventory system used for constructional elements of ship 17.

Element type	Code	Commentary
Planks	S01 – S-- P01 – P--	Numbering started from the keel.
Supports and bracing timbers	F01 S/P	Numbering was started at F10 as excavations proceeded from the bow towards the stern.
Through-beams	B1 S/P	
Stanchions	E1 S/P	
Specific elements or those with undetermined destination	D01 S/P/C	Pieces found in the region of the keel are identified by the letter 'C' (centre).

19 See tables 2.5 and 3.1.
20 Sample reference numbers 03_2010 to 09_2010 (7 samples) and 19_2010 to 22_2010 (four samples).
21 Further wood samples were obtained alongside pieces of ropes and luting: see Chapter Three, table 3.1.

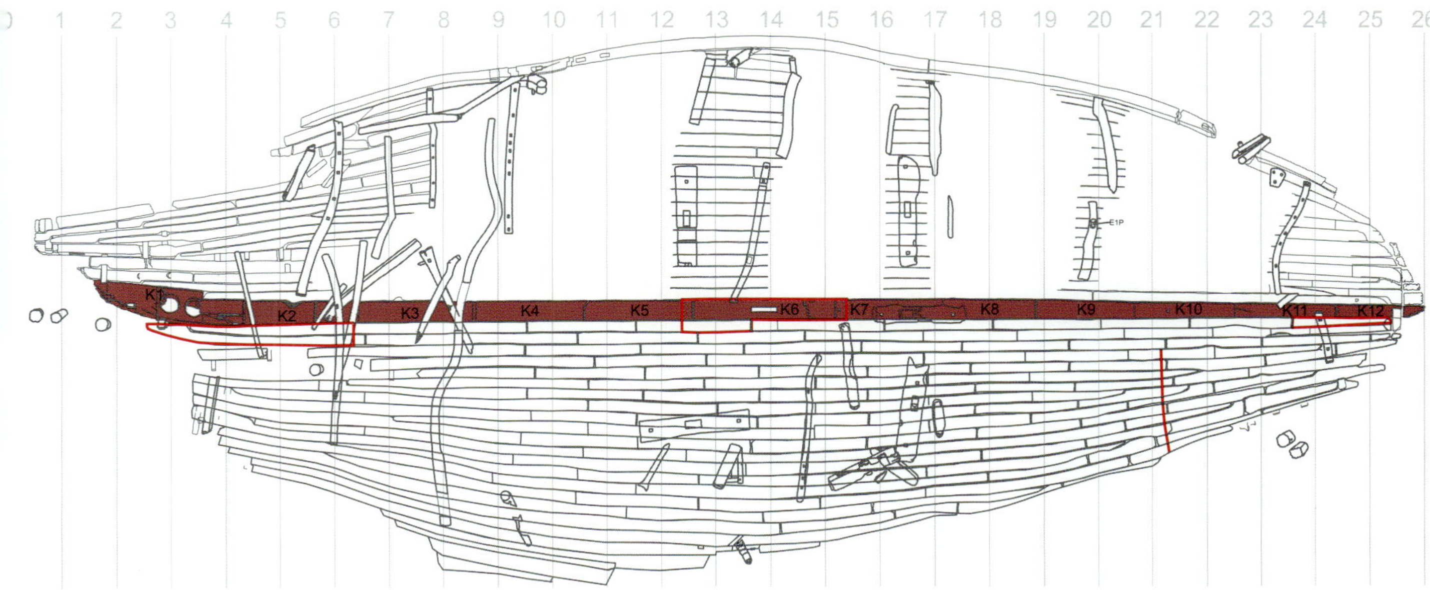

Figure 2.6 Position of the sections of the hull during 2010 (section at the stern) and 2011 excavation campaign. (Drawing P. Sandrin/A. Belov; © F. Goddio/IEASM.)

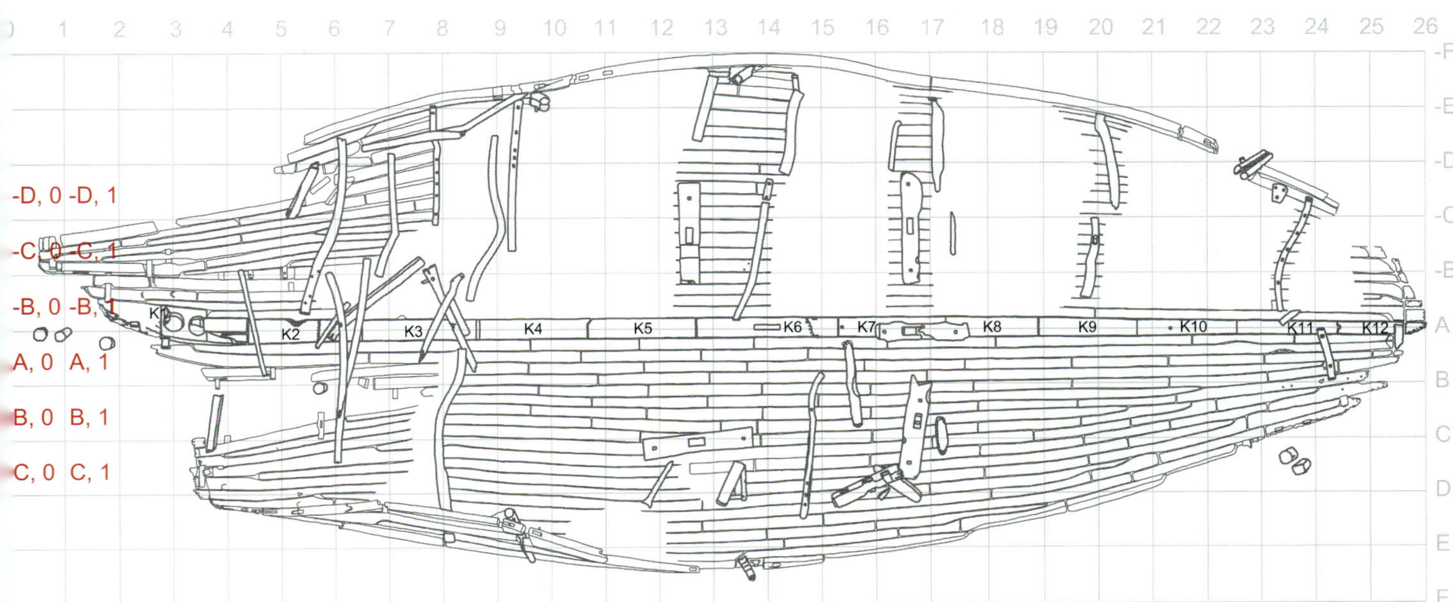

Figure 2.7 Grid system in use during the excavation of ship 17. (Drawing P. Sandrin/A. Belov; © F. Goddio/IEASM.)

complete photo coverages of the wreck to be obtained from which to construct an accurate photomosaic.

Coordinate system

In 2009 the numbering of the base line started at the preserved end of the bow. During subsequent excavations the 'zero' was moved to a point situated along the axis of the keel 1.60 m northeast from the preserved extremity of the hull to its aft. Objects were positioned on the X-axis by positive values from point zero towards the bow; and the Y-axis measures the distance from central line of the keel. Starboard had positive values and port negative ones. The grid system is shown on Figure 2.7. The elevation (Z-axis) was measured from the upper surface of the central segment of the keel (K6).

Preservation

The general state of preservation of the wood was very good, since it was protected from wood-boring molluscs by the dense Nile clays that buried it (Table 2.2). Those parts of the ship that protruded from the clay were heavily degraded. Some parts of the hull—for example,

Table 2.2 Type and number of the preserved pieces of internal construction on ship 17.

Element of Construction	Port	Starboard	Centre
Support	2	2	1
Bracing timber	6	4	
Through-Beam	3	1	1
Stanchion	1	1	2
Hatch		1	
Undetermined	7	2	5

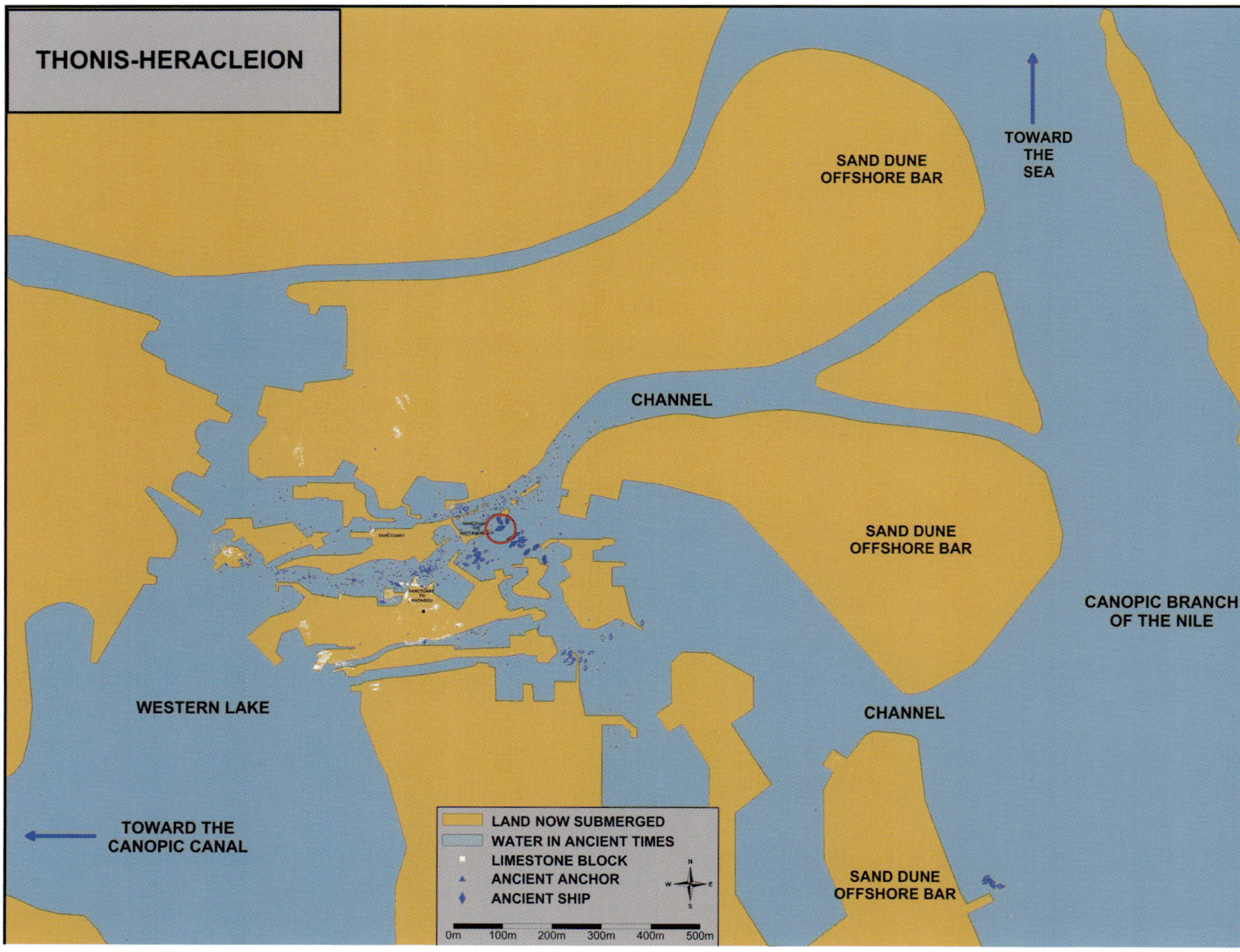

Figure 2.8 Position of ship 17 on the general plan of Thonis-Heracleion. (Map: F. Goddio; © F. Goddio/IEASM.)

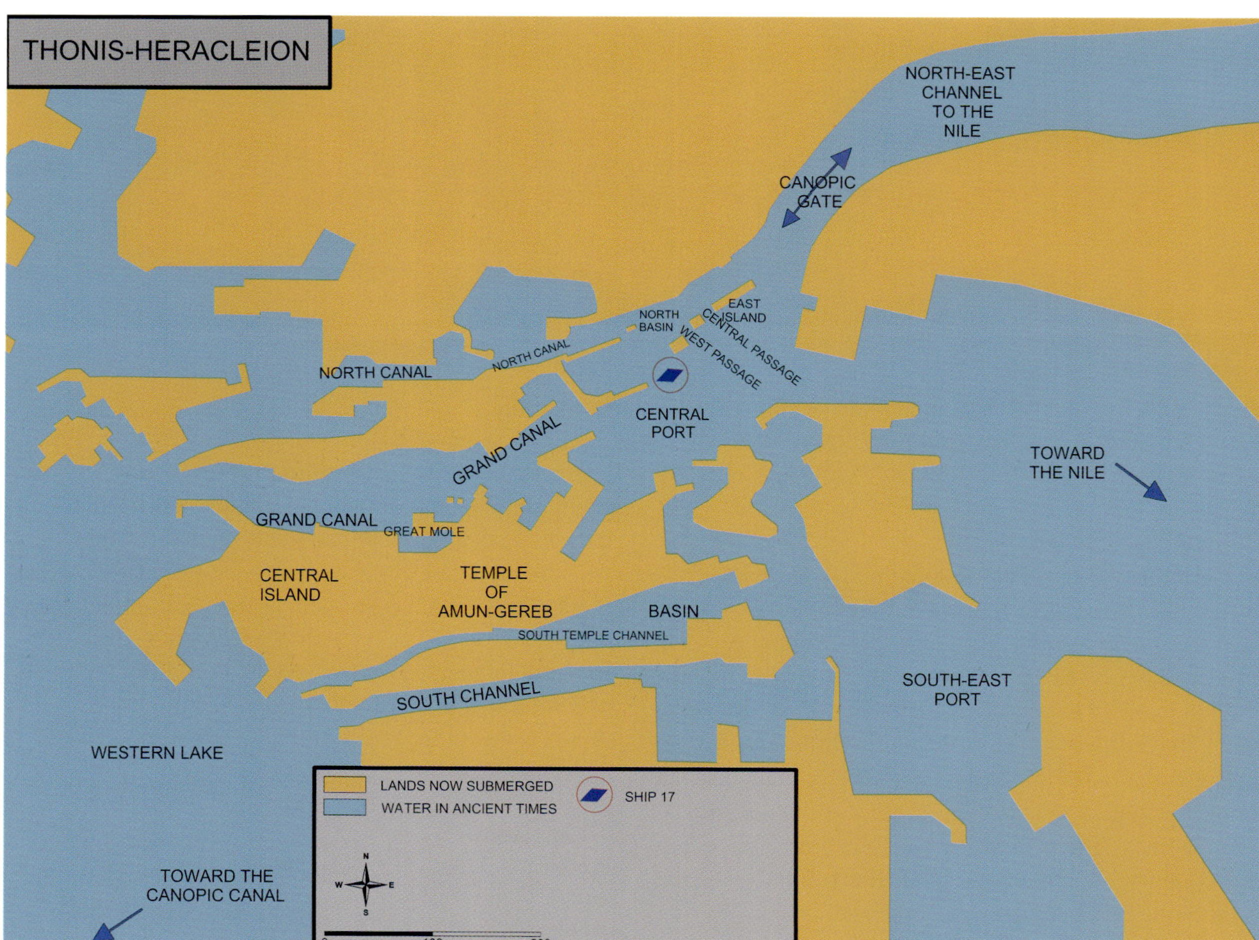

Figure 2.9 Position of ship 17 in the passage between two port basins of Thonis-Heracleion. (Map: F. Goddio; © F. Goddio/IEASM.)

Figure 2.10 The butt ends of the stem and of the first strakes of ship 63 (bottom left) against the broken bulwark of ship 17 (right). (Photo: C. Gerigk; © F. Goddio/IEASM.)

the planking of the starboard at the stern—were also badly damaged owing to post-depositional deformation of the hull. These areas, as well as loose elements of the internal structure of the ship, were secured with sand-filled bags. Nails were used to secure smaller constructional pieces. At the end of each excavation season the ship was completely reburied under sand with the aim of preserving it *in situ*.

The position and context of ship 17

The location of ship 17

Ship 17 was discovered in the passage leading from the Central Port to the North Basin of the city (Figures 2.8–9). Numerous blocks of limestone, some of them four metres long, were oriented to the northeast. They correspond to the remains of substantial buildings and other structures built on and around the islands, outlined in yellow on Figure 2.9.[22]

Eleven ancient ships were discovered about one hundred metres from ship 17. Ship 63 was found near its starboard. The stem of ship 63, which has the same type of construction, broke the upper preserved strakes of ship 17 (Figure 2.10).[23]

The stratigraphic location of ship 17

Ship 17 was found at the depth of 7–8 m below sea level under a layer of sand that was 30 cm thick over the central part of the keel and 40–50 cm thick over the extremities of the hull (layer 3). The layer of sand reached its maximum thickness of 70 cm over the keel at 21 m along the baseline. The extremities of the ship are sealed by a layer of dense grey clay 50–70 cm thick (layer 2). The deepest parts in the centre of the vessel, between 10 and 18 m from the zero point, were under about 1.05 m of sediment in total. In general, the ship had no pronounced heel or pitch; however, the weight of post-depositional sediment caused the hull to collapse outwards along the transversal axis.

A thin layer of clay, two to three cm thick, covered the interior of the planking (layer 1). This layer was enriched by chips of wood (one to five cm long), limestone pebbles, and several sherds of pottery. A layer of fine white clay no more than one cm thick was observed in several limited areas inside the hull near the bottom strakes (layer 1a). Another isolated layer of fine grey sand, 40–50 cm thick, was recorded on the starboard side of ship at 10–11 m from the baseline under the layer of grey clay (layer 1b). This and the underlying layer of clay contained fragments of fish bones (Nile catfish *Claiidae sp.* and Nile perch *Lates niloticus sp.*) and fresh/brackish water shells.[24]

22 Goddio 2007: 111.

23 The radiocarbon date of ship 63 remains undetermined.

24 Ikram 2005 and see also Bewer and Friedman 1989: 60–61, 70–75.

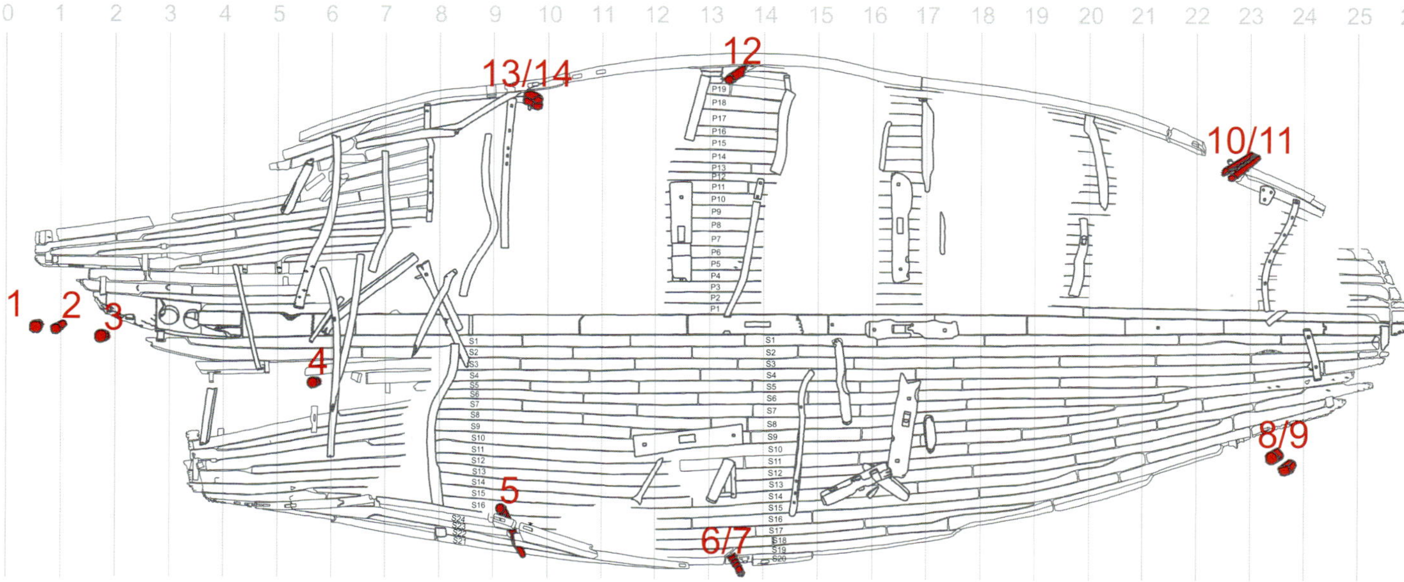

Figure 2.11 Wooden poles discovered around ship 17. (Drawing: P. Sandrin/A. Belov; © F. Goddio/IEASM.)

Figure 2.12 Pole no. 12 viewed from the outside of the hull. In the foreground is the edge of the last preserved strake P23. (Photo: C. Gerigk; © F. Goddio/IEASM.).

The poles

Fourteen substantial wooden poles surrounded ship 17 (Figure 2.11). There were four pairs of poles, which were assembled with transverse tenons and six single poles. Eight poles were circular in section, five were rectangular and one was square. The section of the rectangular and square poles varied from 9 × 15 cm to 15 × 15 cm, while the diameter of the circular ones ranged from 12 to 25 cm. Four poles at the stern were almost vertical; the majority of the others were inclined inside the hull. Except for three poles at the stern (numbers one to three) and numbers eight and nine, the poles passed through the planking, in some cases inflicting serious damage to the hull. Thus, pole 12 broke three strakes and lead to the appearance of a large breach in the planking

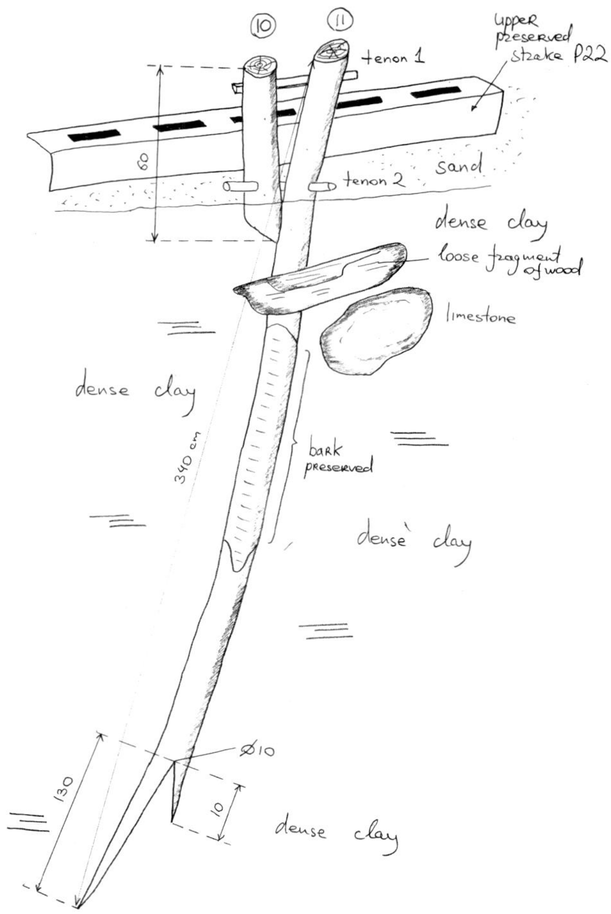

Figure 2.13 Sketch showing poles 10 and 11 uncovered by the sounding. (Drawing: A. Belov; © IEASM.)

(Figure 2.12). It is likely that this change occurred during the post-depositional separation of the hull as it lost its structural integrity and became 'flattened'. The poles appear to have found their way through the planks as part of this process, causing significant breaches and damage. They were not part of a separate placement strategy designed to hold the sunken hulk in place.[25]

The wood sample from pole 3 was identified as cedar (*Cedrus libani*).[26] Employment of this noble wood for a pole may reflect the urgency of the scuttling or simply the lack of other available timber. The preliminary identification of the wood sample from pole five seems to indicate acacia (*Acacia nilotica*).

A sounding was made around poles 10 and 11 in an attempt to understand their deposition and form. Pole 10 was a fragment of wood 60 cm long fastened by a transversal tenon to pole 11, which was itself very long—3.40 m—and had a forked end possibly to anchor it to the sediment (Figure 2.13). It may be inferred that the three other pairs of poles had the same type of construction.

The interpretation of the sequence of deposition of ship 17

Ship 17 was positioned adjacent to, and at exactly the same angle as, the structure to the south-west that seems to be an ancient jetty. This suggests that ship 17 was hulked and reused to extend the jetty into deeper water to allow ships of greater draft to berth, or to provide more space to enable more ships to tie up on the jetty. The placement of the poles around the hull suggests that the ship was floated into place, tied to the stakes so that it did not move and then probably had infrastructure, possibly the top of the jetty, built over it while still floating.[27] The presence of fine clay, sand and molluscs inside the hull may be explained by infiltration through small breaches that could appear at the time. The sinking of the ship and the decommissioning of the elongated jetty dates to a later period. It will be shown below that the sinking of ship 17 probably happened in the first half of the fifth century BC or a little earlier.[28] The collision with ship 63 that damaged ship 17 probably also dates to this period, although it should be noted that this damage was probably insufficient to cause the hulk of ship 17 and its jetty to sink.

State of preservation of the hull

The preserved part of the hull is 25.2 m long and 9.4 m wide. The general integrity of the hull was very good. The stern planks of the upper starboard, strakes S21, S22 and S23, were broken and inclined inside the hull owing to the impact of the stem of ship 63. The accompanying segment of the keel, K1, also had a pronounced heel to port (43.5° from the horizon). This would suggest that the sinking of ship 63 damaged ship 17 and perhaps also the infrastructure of which it was a part.

In general, the port side of the vessel was better preserved than the starboard and included more pieces of internal structure *in situ*. The ship experienced some deformation of the wooden hull during its decomposition; flattening and opening.[29] The upper surface of the central part of the keel had almost the same elevation as the preserved extremities, and the stern and bow of the ship form two almost symmetrical curves in longitudinal profile (Figure 2.14).

Considerable opening of the hull was apparent in transversal axis (Figure 2.15), but despite the important deformation of the hull, there were not many fractures in the planking. This can be explained by the short length of individual planks that makes the overall planking rather flexible, especially in the absence of

25 See Richards 2008: chapter 6.
26 Analysed by the Laboratoire de Datation par le Radiocarbon of the IFAO in 2009. Identification by A. Fahmy, Helwan University, upon the Laboratory's request. Analysis report IFAO 371.
27 Pers. comm. Robinson, 7th November 2016.
28 But see Grataloup below.
29 The principal stages of the decomposition of a wooden hull are described in Muckelroy 1976. Associated erosion processes are considered by Quinn 2006.

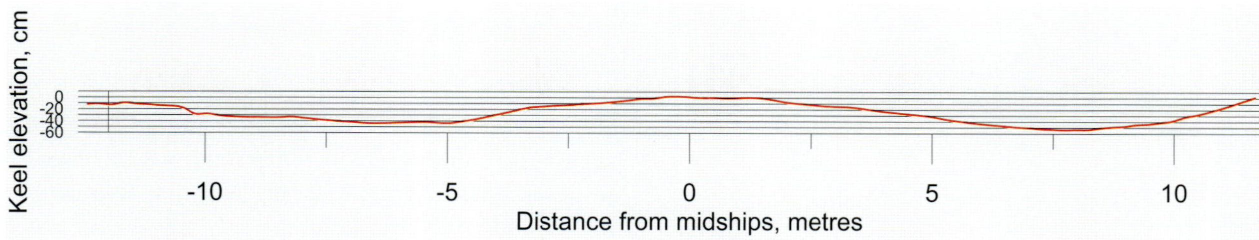

Figure 2.14 *In situ* profile of the keel. Horizontal and vertical axes are in the same scale. (Drawing: A. Belov.)

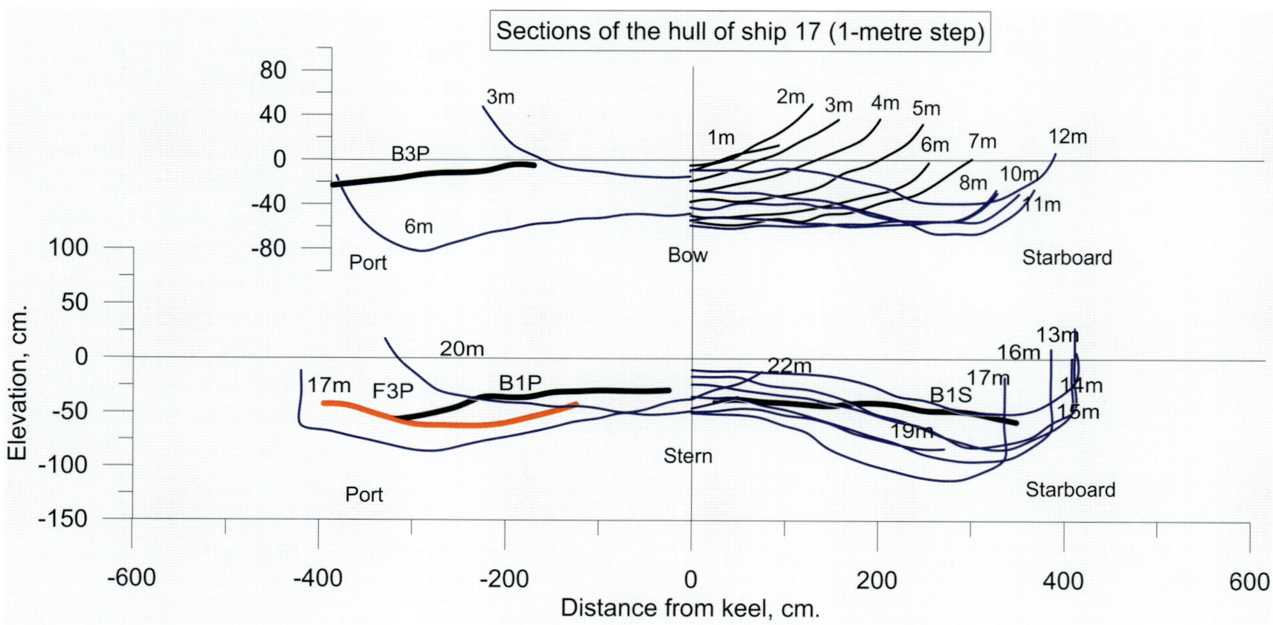

Figure 2.15 Transversal view of ship 17 *in situ*, starboard sections are shown at each metre along the length of the keel. (Drawing: A. Belov.)

any developed internal structure. There are, nonetheless, several zones where the fractures were pronounced (Figure 2.16). Furthermore, while the planks themselves were held together by long internal tenon-ribs, where visible the majority of these were also fractured in a like manner.

It is interesting to note that the fracture of keel segment K6 was situated precisely in the middle of the hull, most probably owing to the descent of the ends of the ship. The preserved part of the keel included 12 segments. The stern was considerably eroded resulting in the incomplete preservation of the first element of the keel (K1). The same is true for the last preserved segment (K12) at the bow. In total, 24 strakes were preserved on the starboard and 22 on the port. The number and distribution of the preserved pieces of internal construction are shown in Table 2.2 and Figure 2.17.

Objects from ship 17

The excavation of ship 17 revealed little in the way of material culture that could be directly associated with the vessel, indicating that it was clearly not a shipwreck. Nevertheless, several complete ceramic vessels and sherds were discovered in the clay in association with the planking or under some of the ship's structure, suggesting perhaps that they were *in situ* when the hulk sank. More archaeological material was recovered from the sand layer overlying the vessel. As these layers—and the objects within them—are relatively mobile, it is not possible to link any of them to the ship.

A deformed spherical lead container was discovered near keel segment K1 at the stern (2, B) (Figure 2.18). It was initially about 40 cm wide, with a rim diameter of 20 cm, and its walls were 5 mm thick. The container was found among fragments of wood belonging to ship 17,

Table 2.3 Artefacts found in the layer of sand over ship 17.

Artefact	Inv.Num.	Dimensions	Position
Rectangular limestone basin	L1_11263	19 × 19 × 5 cm	(14, -A), under 20 cm of sand
Circular limestone basin	L1_11262	∅ 9 cm, height 11 cm	(14, -A), under 20 cm of sand
Fragment of terracotta figurine (joint legs of a woman)	L1_11248	10 × 5 × 3 cm	(14, -A), under 30 cm of sand
Weight, terracotta	L1_11247	5.5 × 4.5 × 4.5 cm	(14, B), under 20 cm of sand
Head of limestone figurine	L1_11493	3 × 2.5 × 2.5 cm	(7, B) under 40 cm of sand

Figure 2.30a–b Fragments of an amphora close to Milesian or related production of the period *c.* 450–425 BC (L1.12056) *in situ*. (Photo: A. Belov; © F. Goddio/IEASM.)

probably relates to a shape known in the Delta from contexts that date from the end of the fifth century BC to the first half of the fourth century BC.[67]

Two types of amphora sherds were found in the clay adjacent to the planking of ship 17. The first was the neck, rim and double handle fragment of a probably Milesian or related amphora dated to *c.* 450–425 BC or slightly later (Figure 2.30a–b).[68] The origin of the sherds belonging to another amphora's rim is not certain, however, these seem to date to the middle of the fifth century BC.[69]

On the whole, the ceramic assemblage excavated from contexts above and related to ship 17 are coherent in terms of both their date (Table 2.4) and range of forms when compared to similarly dated contexts elsewhere in Thonis-Heracleion. While the ceramics in the overlying

Table 2.4 Dates for ceramics from ship 17. Finds from upper layer of sand are shown in orange and those from underlying layer of dense clay in blue.

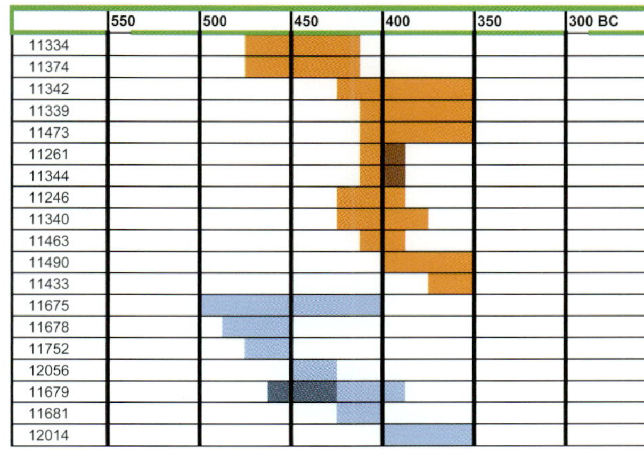

Table 2.5 Radiocarbon dates for wood samples from ship 17.

Sample	IFAO -143 (planking)	IFAO-144 (tenon of the planking)
Conventional date	2493 ± 46 BP	2536 ± 46 BP
Calibrated dates	cal 786-486 BC (89%)	cal 804-518 (95,4%)
	cal 694-540 BC (53,2%)	cal 794-748 BC (23,9%)
	cal 766-721 BC (15%)	cal 644-589 BC (23,6%)
		cal 688-665 BC (11,7%)

See http://www.ifao.egnet.net/c14 (last consulted 23 August 2015).

mobile layer of sand, and also perhaps those recovered from the dense clay next to the hull, were not associated with the active life of ship 17, they do provide evidence to suggest a date for its abandonment. Both the incorporated and overlying ceramics point towards a time in the first half of the fifth century BC—or perhaps a little earlier—for this event.

The date of ship 17

Radiocarbon date

The date for the construction of the ship was established from samples subjected to radiocarbon dating. The results are presented in Table 2.5.[70]

Conclusions

The date of ship 17 is problematic as neither cargo nor the crew's belongings were found on board. Radiocarbon calibrated dates range from 804 to 416 cal BC at 2 sigmas (probability of 95 per cent). Consequently, it is the ceramic evidence found in the clay layer immediately adjacent to the inner surface of the planking and preserved *in situ* under fallen timbers inside the hull that narrows down

67 Defernez 2003: 317, types 198, 199 and pl. LXXI, type 203.
68 Grid 9, -E near the through-beam B1P and Inv. no. L1.12056.
69 L1.11752.

70 Format of presentation as per 'Radiocarbon' journal. See http://www.radiocarbon.org/Authors/author-info.pdf.

the radiocarbon dates. Seven vessels form a coherent group that suggests the date in the first half of the fifth century BC as the time when the hull finally sinks. Taking into consideration the material from the upper layer of sand *terminus ante quem* for the site formation can be placed in the middle of the fourth century BC.

Defining the *terminus post quem* for the ship is more problematic. Calibrated radiocarbon dates leave a wide range of 794–540 BC at 1σ (probability of 68 per cent). It is also necessary to keep in mind that according to contracts of the lease-sale of ships (*misthoprasia*) in Roman Egypt the longevity of ships could reach 50–60 years.[71] Furthermore, it is likely that ship 17 was most probably reused at the end of its sailing life.

Summing up the available dating evidence from radiocarbon and ceramics, it would appear that ship 17 was an old vessel that was hulked most likely sometime in the sixth century and reused as a floating jetty that finally sank during the first half of the fifth century BC or a little earlier.

71 Arnaud 2012: 95.

II

The construction of ship 17

3 The species of wood used

Identification of the wood

Twenty-seven wood samples were taken from the ship for wood analysis; of these, eight were analysed by the Radiocarbon Laboratory of the French Institute (IFAO) in Cairo and a further 19 examined by the author (Table 3.1). The samples were taken from both the hull—keel planks and tenons—and the through-beams and support structures to get as comprehensive an overview as possible of the different types of wood that was used to construct the vessel.

While the initial results from samples taken in 2007 suggested that the planking was made of Holm oak (*Quercus ilex*) or Kermes oak (*Q. coccifera*),[1] further analyses on samples taken in 2009–2011 would suggest that ship 17 was built overwhelmingly from different species of acacia. According to Theophrastus (repeated by Pliny the Elder), the Egyptians were able to cut 12-cubit planks of acacia (about 6.30 m).[2] Acacia wood, and especially *Acacia nilotica* (*sndt*), was widely used in ancient Egyptian boatbuilding in conjunction with other local species.[3] In the Old Kingdom

Table 3.1 Results of the identification of wood from ship 17.

Sample No. (IEASM)	Sample Description	Laboratory	Report No.	Wood species
	2007	IFAO		
H 2003 N6	Plank		IFAO_100 (143)	*Quercus* sp. (evergreen species)
H 2003 N7	Tenon of the planking		IFAO_101 (144)	*Acacia* cf *nilotica*
	2009	IFAO		
7_2009	Pole 03		IFAO_371	*Cedrus libani Loud*
9_2009	Plank		IFAO_373	*Acacia raddiana Savi.*
10_2009	Tenon of the planking		IFAO_374	*Acacia* sp.
11_2009	Support F5P		IFAO_375	*Acacia nilotica* (L.) *Delile*
12_2009	Bracing timber F6P		IFAO_376	*Acacia* sp.
13_2009	Through-beam B2P		IFAO_377	*Acacia raddiana Savi.*
	2010		Preliminary determinations by the Author	
03_2010	Support F8S			*Acacia nilotica*
04_2010	Through-beam B1S			*Acacia* sp.
05_2010	Support F5S			*Acacia* sp.
06_2010	Through-beam B1P			*Acacia* sp.
07_2010	Pole 06			*Acacia nilotica*
08_2010	Pole 07			*Acacia nilotica*
09_2010	Bracing timber F10P			*Acacia* sp.
19_2010	Keel segment K2			*Acacia* sp.
20_2010	Tenon (keel segment K2- strake P1)			?
21_2010	Keel segment K1 (with openings of the steering system)			*Acacia* sp.
22_2010	Tenon (keel segment K3- strake P1)			*Acacia* sp.
	2011		Preliminary determinations by the Author	
01_2011	Tenon peg			*Acacia* sp.
02_2011	Wedge			*Acacia* sp.
03_2011	Wedge of tenon N1 from keel segment K6 (port)			*Acacia* sp.
04_2011	Wedge of tenon N1 (keel segment K6)			*Acacia* sp.
05_2011	Tenon N1 (keel segment K6)			?
06_2011	Tenon N4 (keel segment K6)			?
07_2011	Keel's segment K6			*Acacia* sp.
22_2011	Strake S1			*Acacia nilotica*

1 Analysed by the Laboratoire de Datation par le Radiocarbon of the IFAO in 2007. Identification by A. Fahmy, Helwan University, upon the Laboratory's request. Analysis report IFAO 143. See http:/www.ifao.egnet/c14 (last consulted 15 December 2015).

2 Theophrastus *Historia Plantarum* 4.2.8; Pliny the Elder *Naturalis Historia* 13.63.19.

3 Ward 2000: 15–16.

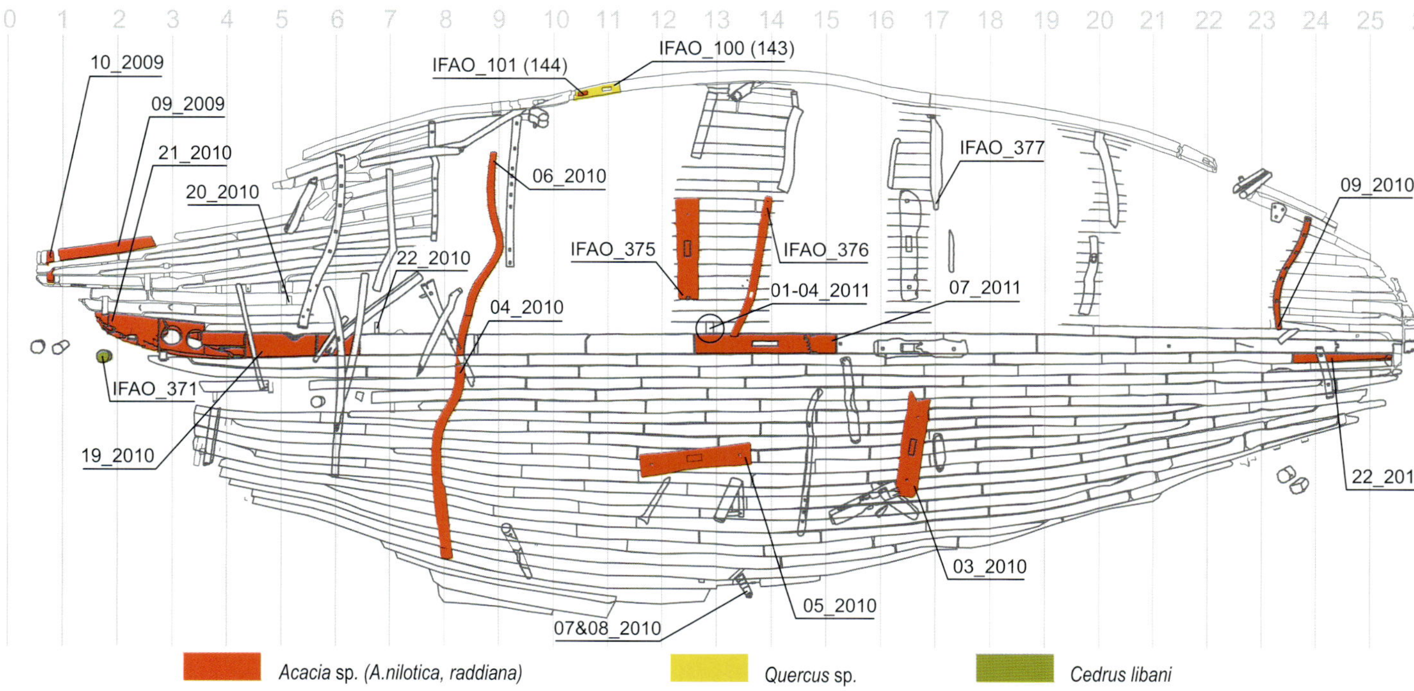

Figure 3.1 Plan of distribution of wood species used in the construction of ship 17. (Drawing: P. Sandrin/A. Belov; © F. Goddio/IEASM.)

Figure 3.2 First strake of the starboard in section showing the heartwood (right). (Photo: C. Gerigk; © F. Goddio/IEASM.)

(*c*. 2543–2120 BC), Nile freighters were built of acacia planks that were about 32 m long.[4] Acacia was identified as the construction material in the boats from Lisht (Middle Kingdom, *c*. 1950 BC).[5] Traditional boats of the Upper Nile continued to be built of *Acacia nilotica* into the late 1930s.[6]

4 Boats *sḥt* and *sȝṯ* built by general Weni (*wni*) during the rule of Pepi I (sixth dynasty, 2232–2283 BC).

5 Together with tamarisk, see: Ward 2004: 15.

6 Hornell 1943a: 28: 'Timber—invariably local acacias, usually sunt (*Acacia nilotica*), an extremely hard wood but brittle and troublesome to work. It is difficult to obtain in long running lengths, so, apart from the keel plank, the hull is built up of comparatively short lengths of thick planking halved together in the strakes.'

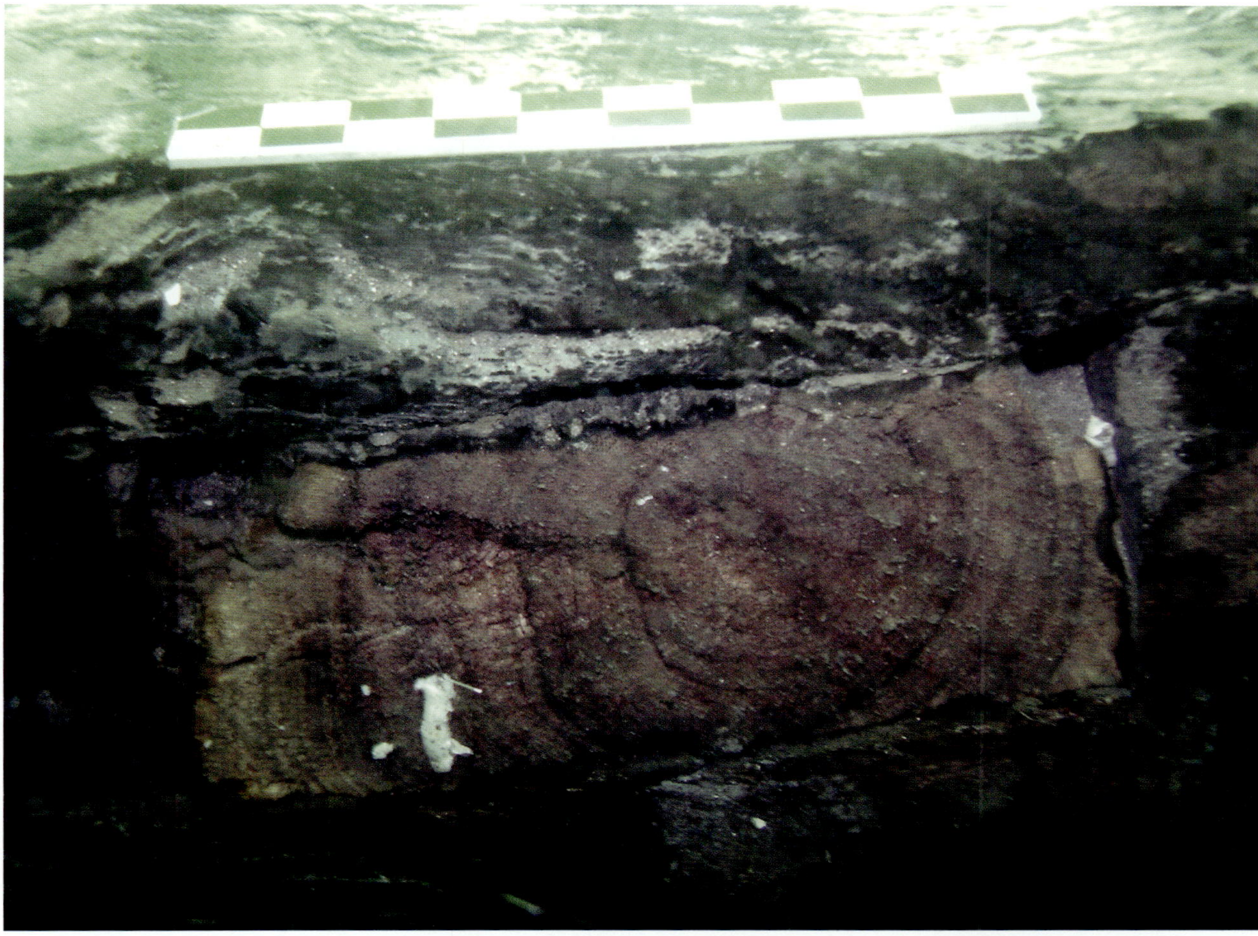

Figure 3.3 Typical appearance in section of a tenon of the planking. (Photo: A. Belov; © F. Goddio/IEASM.)

The wood of the acacia tree is very strong, tough and heavy. Its average density is between 650–830 kg/m^3 at 15 per cent moisture content.[7] In addition it is brittle, while its high silica content rapidly blunts cutting tools. One great advantage for shipbuilding is that this wood is very durable if seasoned well and does not rot in water.

As seen in Figure 3.1, acacia wood was the main construction material for ship 17. It was identified in the samples taken from the planking, keel, bracing timbers, supports, and joints. Acacia also dominates as the constructional material for the majority of the ships from Thonis-Heracleion. Preliminary xylological analysis showed that among 63 ships about 80 per cent have planking made of acacia, which is a strong argument for their local origin.[8]

Sawing of wood

Due to the relatively short excavation seasons and the ship remaining *in situ* and needing to rebury it to protect it, there was insufficient time available for a detailed dendromorphological analysis. Nevertheless, the following preliminary observations may be suggested.

Generally the wood was used very frugally and many planks contain knots and fissures.[9] The keel segments and the planks of ship 17 were cut from squared logs of acacia that contain the heartwood (Figure 3.2). The inboard surfaces of the planks appear to be flat, while their outboard surface is sometimes slightly curved. This latter aspect of the planking resembles that seen in the Dahshur boats.[10] The outer surface of the keel, however, was flattened with an adze. The tenons of the planking were flat cut, with the heartwood invariably present (Figure 3.3).

The bracing timbers have different shapes in cross-section with octagonal being the most common. Through-beams were shaped only at their ends where they were joined to the planking; otherwise they are cut from naturally-curved branches and trunks that were stripped of bark.

7 Fagg and Mugedo 2005: 'At 15% moisture content, the modulus of rupture is 88 N/mm^2, modulus of elasticity 11,060 N/mm^2 and the compression parallel to grain 52.5 N/mm^2'.

8 Fabre and Belov 2012: 109–110.

9 Cf. Ward 2004: 14. 'Freight boats depended on flatter and shorter planks and frames frugally sawn from trimmed trunks, seemingly in standardized shapes and sizes that took advantage of the tree's natural curvature'.

10 Ward 2000: 87.

Figure 3.4 Traces of an adze on the outer surface of keel segment K6. (Photo: A. Belov; © F. Goddio/IEASM.)

Figure 3.5 Cutting of the outer surface of a keel with an adze. Tomb of Ti in Saqqara (fifth dynasty, 2494–2345 BC). (Photo: A. Belov; © F. Goddio/IEASM.)

Traces of tools

Traces of an adze were distinctly visible on the outer surface of the keel, spaced at 1.0–1.5 cm (Figure 3.4) and it is noteworthy that this method of construction is mirrored on Old Kingdom representations showing the same activity (Figure 3.5). Tool marks were well-preserved in the mortise holes cut through the planking. It is clear that they were cut from both sides of the plank, meeting somewhere in the middle, with a chisel with a

Figure 3.6 Traces of a chisel inside the channels of keel segment K6. (Photo: A. Belov; © F. Goddio/IEASM.)

blade *c.* 2 cm wide. It is likely that the planks were initially cut and then the mortises were chiselled into them at a stage prior to final assembly. This will be discussed further below (Figure 3.6).[11]

The two shafts for the steering oar of the ship were cut directly into the keel (segment K1) with a small chisel with a blade *c.* 1 cm wide.

11 See Chapter 9.

4 Longitudinal structure

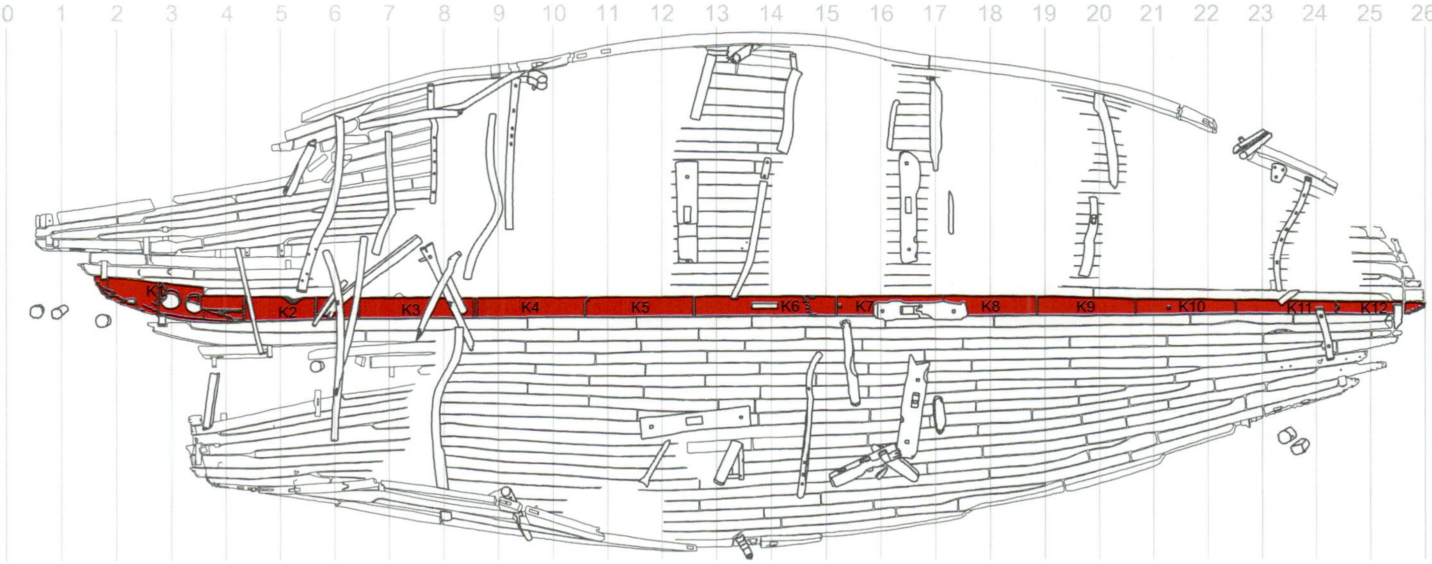

Figure 4.1 Plan of the keel of ship 17. (Drawing: P. Sandrin/A. Belov; © F. Goddio/IEASM.)

Form and dimensions of the keel

The surviving keel of ship 17 is 24.2 m long (Figure 4.1). It is made of 12 short segments of squared acacia (*Acacia nilotica, A. raddiana*) logs, with the first and last (K1 and K12) being heavily damaged. There are two openings in segment K1, at the stern of the ship, through which the steering oar(s) of the vessel would have passed. In the middle of the central segment K6 there was a mortise for the spur of the mast.

The length of the keel segments varied between 1.62 and 3.05 m, with an average of 2.02 m. Starting from the middle of the hull, the segments of the stern part were longer. The average width of the segments was also greatest at the stern (56 cm) and smallest at the bow (21 cm), while the central part of the hull was characterised by a constant width of about 35 cm (Figure 4.2). The keel was considerably wider than the strakes along the entire length of the hull (Figure 4.3). The keel was mostly rectangular in section (Figure 4.4) and reached its maximum thickness in the middle of the hull where it was almost square. The lower surface of the keel was flush with the outer planking and the keel projected up inside the hull. This elevation was minimal at the preserved extremities of the ship (20 mm) and reached 130 mm at the middle of the hull.

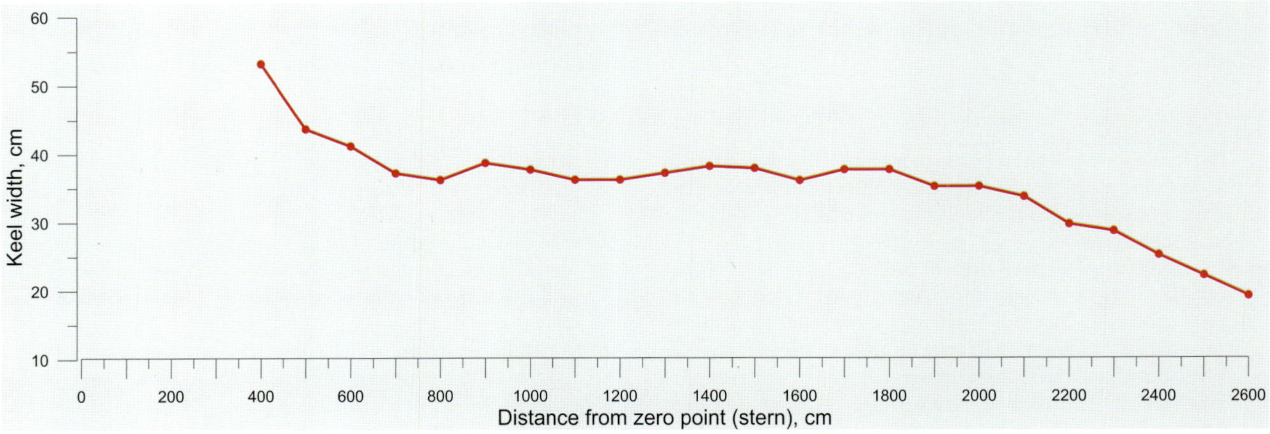

Figure 4.2 Width of the keel segments along the hull.

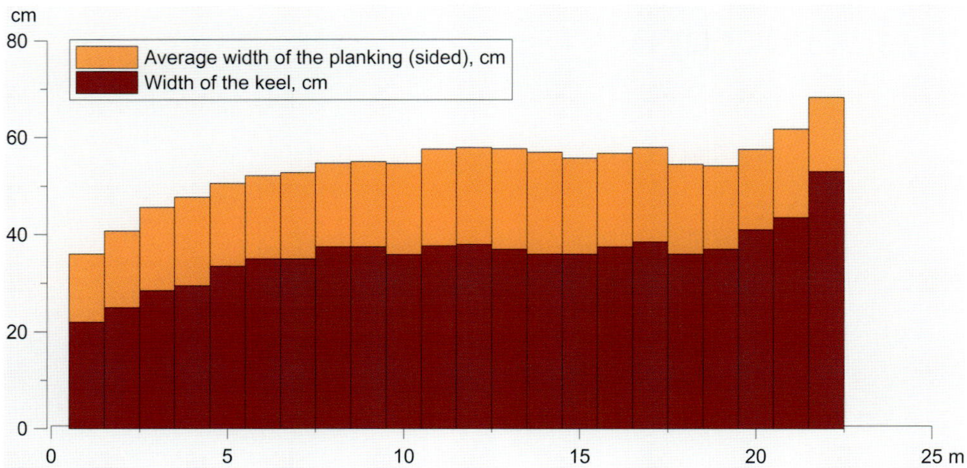

Figure 4.3 Width of the keel segments plotted against the average width of the planking along the hull.

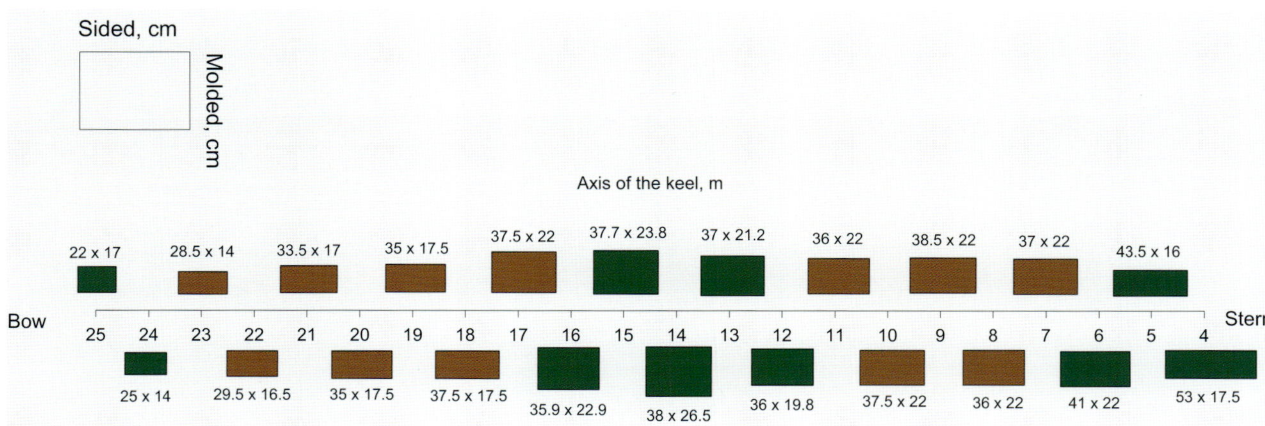

Figure 4.4 Shape and size (in cm) of the section of the keel. Attested sections are shown in green colour while theoretical ones are in brown.

Stem and stern timbers[1]

The keel would have ended with massive stem and stern timbers that were not preserved in ship 17 but are attested from other vessels of the Late Period from Thonis-Heracleion. It is possible that segment K1 is the final piece of the keel at the stern, as suggested by the converging triangular form of the piece and by the presence of the openings belonging to the steering system of the ship. K1 was not, however, complete and the remaining port strakes exceed its length by 1.08 m, indicating that the existence of a short additional segment, or a continuation of K1, cannot be completely excluded.

The remaining section of K12 contained no information about how the hull of ship 17 was closed at the bow.[2] Fortunately, a parallel can be drawn between ship 17 and its neighbour some 100 m to the south: ship 43, which has been radiocarbon-dated to the period between early eighth and early fifth centuries BC.[3] The construction of these two ships appears to be very similar[4] and at the bow of ship the keel ends with a large piece probably carved from acacia wood (Figure 4.5). This piece, installed as a direct continuation of the keel, was characterised by a triangular shape in plan and, probably, that of a half cone in volume. Its function was to close the crescent-shaped hull by receiving strakes with varying angles of entry, from the flat bottom strakes to the quasi-upright bulwark.[5]

1 The terms stem and stern posts are intentionally avoided as the strakes were assembled from the extremities to the last segments of the crescent-shaped keel in accordance with ancient Egyptian boatbuilding tradition.

2 It is interesting to note that in addition to general ancient Egyptian terms of *w3ḏt* and *ḥ3t* ('stem', 'bow of ship', 'forepart', 'prow of boat', see Jones 1988: 159, 173, 3.32 and 3.97), there existed some more specific ones. The term *m3-ḥ3wty* (ibid: 166, 3.62) signified 'bowpiece' and *ibs(t)* (ibid: 153, 3.7) may be understood as 'bow-timbers'. The word *tpty* that was interpreted as 'bow head, stern head' is even more figurative (Barguet 1986: 622 and Jones 1988: 192, 3.174).

3 IFAO 785–481 BC (89 per cent). Analysed by the Laboratoire de Datation par le Radiocarbon of the IFAO in 2007. Analysis reports IFAO 54 and IFAO 55.

4 The major similarities include the same wood used in its construction, the same size of the planks, the same planking plan and method of assembly, and steering system. See Robinson 2015.

5 A parallel with papyrus rafts, which is very prominent for some aspects of Egyptian shipbuilding, seems to be pertinent in this case, as these pieces bunch the strakes in a similar way that bundles of papyrus are tied together at the ends of a raft. For raft construction, see Landström 1970: 17–9.

Figure 4.5 Massive stem timber at the bow of ship 43 from Thonis-Heracleion. (Photo: C. Gerigk; © F. Goddio/IEASM.)

Mortises on the keel

A rectangular mortise is visible in the middle of upper surface of the central keel segment K6, which was probably intended to receive the spur of the mast.[6] Two other mortises were discovered in the sides of the same segment K6.[7] The port one still retained the end of bracing timber F6P (Figure 4.6). Its starboard counterpart probably played the same role. It seems logical that bracing timbers resting on the keel reinforced the middle of the hull in the vicinity of the mast.

Constructional piece D1C, situated above the keel at 16–17.5 m of baseline (see Figure 2.17), was attached to the keel with two tenons that are square in section.

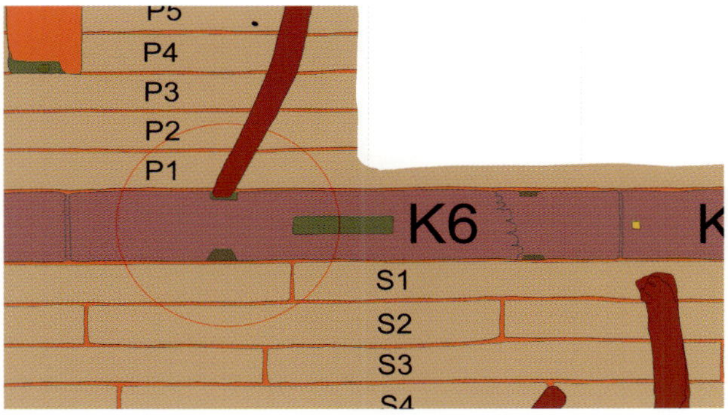

Figure 4.6 Plan of two mortises in the sides of keel segment K6 (red circle). (Drawing: P. Sandrin/A. Belov; © F. Goddio/IEASM.)

Joints between the keel and the planking

The keel and the first strakes of the planking were secured transversally by long tenons, much in the same way as for the other zones of the planking.

The joints between the segments of the keel

The first line of the planking was attached to the keel by long tenons whose mortise holes were cut transversally through the width of the keel. This is discussed in greater detail in Chapter 5. It is important, however, to note at this point that the joints between the segments of the keel and the joints between adjacent planks were staggered and did not overlap. This is clearly what Herodotus meant when he said that Egyptian shipwrights 'cut planks two cubits long and arrange them like bricks'.[8]

The joints between segments of the keel were studied in three areas: the bow, the middle of the hull and the stern (Figures 4.7–4.11 and Table 4.1). Tongue-and-groove joints were used to assemble the segments of the keel. At the bow and the stern, the joint was situated in the lower third of the thickness of the keel whereas in the

6 See Chapter 8, Figures 8.1 and 8.2.

7 Four mortises are shown on Sandrin's plan. However, the two forward 'mortises' of K6 are most probably just gaps connected with transversal fractures due to the deformation of the hull.

8 Herodotus *Historia* 2.96. See also this volume, Chapter 10.

Table 4.1 Dimensions of the keel segments of ship 17 (cm).

Segment	K1	K2	K3	K4	K5	K6	K7	K8	K9	K10	K11	K12	Avg
Length, cm	305.0	209.8	212.9	200.8	197.7	258.8	193.8	168.1	167.5	182.7	169.2	162.9	202.4
Sided_East, cm	56.0	44.6	37.8	36.7	36.7	35.7	35.1	35.1	34.7	34.7	34.7	23.5	37.1
Sided_West, cm	45.6	37.9	36.7	36.4	36.1	36.1	33.7	33.7	23.3	23.3	23.3	21.4	32.3
Sided_Average, cm	51.4	41.3	37.3	36.6	36.4	35.9	34.4	34.4	29.0	29.0	29.0	22.5	34.8
Molded, cm	17.5	16	22		19.8	21.2–26.5	22.9				11.9–14.9	17.0–17.3	19.1

Table 4.2 Dimensions of tongue-and-groove joints between the segments of the keel (mm).

Joint	Groove width	Tongue width	Groove depth	Tongue depth	Position
K1/K2	51	40	56	51	horizontal
K2/K3	62	59	71	41	horizontal
K5/K6	47	47	65	70	horizontal
K6/K7	72	33	61	58	horizontal
K11/K12	44	37	54	52	vertical
Average	55	43	61	54	

middle of the vessel (joint K5/K6) the joint was in the middle of the keel.

With the exception of segment K12, the tongues of the studied joints were always cut at the far end of each segment looking from the centre of the keel. Thus it is probable that the middle segment K5 had two tongues at each end. On average, the grooves were 55 mm wide and 61 mm deep and the tongues 43 mm wide × 54 mm long (Table 4.2). Unlike the other joints studied, the one between segments K11 and K12 in the region of the bow was vertical and reinforced by a key (Figures 4.10–4.11). Traces of luting were visible around this joint.

Discussion[9]

Representations of ancient Egyptian watercraft before the New Kingdom give no hint of the presence of keels in their construction. Likewise, excavated ships show no evidence of the use of a keel: the planking of the Old Kingdom Khufu I boat (*c.* 2566 BC) was built around three bottom strakes, while the Middle Kingdom Dahshur boats (*c.* 1878–1841 BC) had a single central strake only slightly thicker than the rest of the planking.[10] The Late Period Mataria boat (*c.* 450 BC) was also constructed with a keel-plank, which is notably a characteristic feature of *nuggars*, the traditional boats of the Upper Nile.[11] However, all of the above-mentioned vessels are river boats and the absence of a keel is unsurprising as the advantages of flat-bottomed craft with shallow drafts are undeniable in a riverine environment.[12] The ancient Egyptians also built seagoing vessels without keels, as indicated by the ships from the temple of Sahure (2487–2475 BC).[13] Internal girders, hogging trusses and thick planking were a specifically Egyptian means of increasing the structural strength of their crescent-shaped hulls in the absence of the keel.

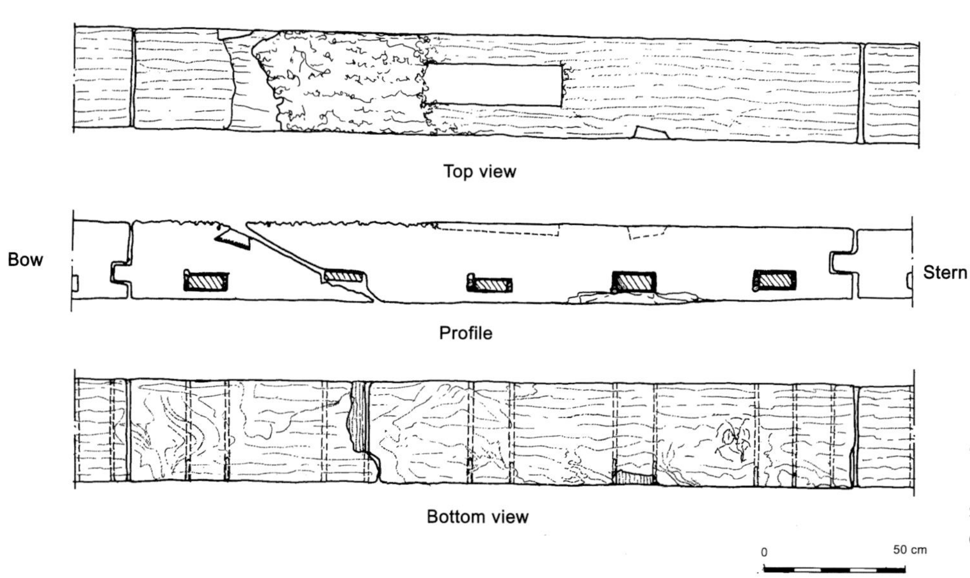

Figure 4.7 Plan and profile of keel segment K6 showing its joints with adjacent segments. (Drawing: P. Sandrin; © F. Goddio/IEASM.)

9 The material of this section was published in IJNA (see Belov 2015).
10 Ward 2000: 85, 97, 102.
11 Clarke 1920: 49; Hornell 1943a: 28.
12 Cf. Cooper 2014 for a vivid account of the frequency with which ships and boats on the Nile regularly grounded.
13 Assmann 1913: 133–66; Landström 1970: 63–7.

Figure 4.8 Lateral view of the joint between segments K2 and K3 (viewed from the starboard side). (Photo: C. Gerigk; © F. Goddio/IEASM.)

Iconographic and epigraphic material dating to the New Kingdom (*c.* 1550–1069 BC) may point to the appearance of a keel, or its analogue, in the construction of Egyptian ships. The representations of the seagoing ships of the Punt expedition of Queen Hatshepsut (1473–1458 BC) from Deir el-Bahari are probably the oldest images that can be included in the discussion. Landström believes that the keel played an important role in the structure of these ships and suggests that it ran along the entire length of the hull.[14] Partridge describes this constructional element as a '...more conventional keel, rather than just a centre plank'.[15] At the same time, many scholars deny that the ships of Hatshepsut had a keel at all. Some, such as Casson[16] and Lloyd,[17] believe that a hogging truss was applied to reinforce the longitudinal structure of these ships precisely because they lacked a keel. Vinson proposes that these ships were flat-bottomed and did not have a true keel. However, he admits that contemporary ship models show a thickened bottom plank that projects at the bow and stern.[18]

Three models from the tomb of Amenhotep II (1427–1400 BC), representing *bik*-ships, possess equally imposing stem and stern posts. These show a direct continuation of keel-like structures that end in high finials. No keel, however, is visible in the middle part of the hull. The models are decorated with the depictions of Montu (*mnṯw*)—the Egyptian god of war—and are thus probably meant to represent royal warships.[19] It seems that

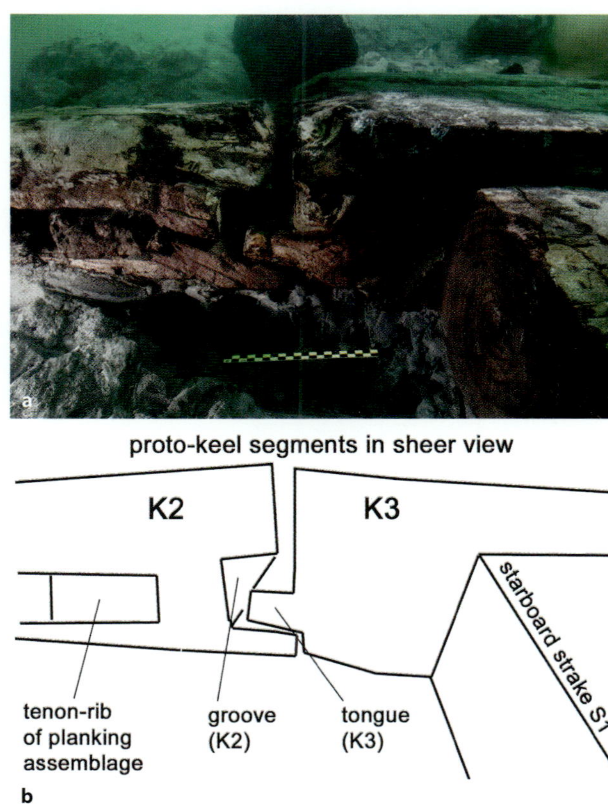

Figure 4.9a–b Lateral view of the joint between segments K2 and K3 (viewed from the starboard side). (Photo: C. Gerigk; drawing: A. Belov; © F. Goddio/IEASM.)

14 Landström 1970: 122–7.
15 Partridge 1996: 54.
16 Casson 1959: 14–5.
17 Lloyd 1972a: 271 note 3.
18 Vinson 1994: 38.
19 Landström 1970: 107–8, figs 338, 339; Werner 1986: 123.

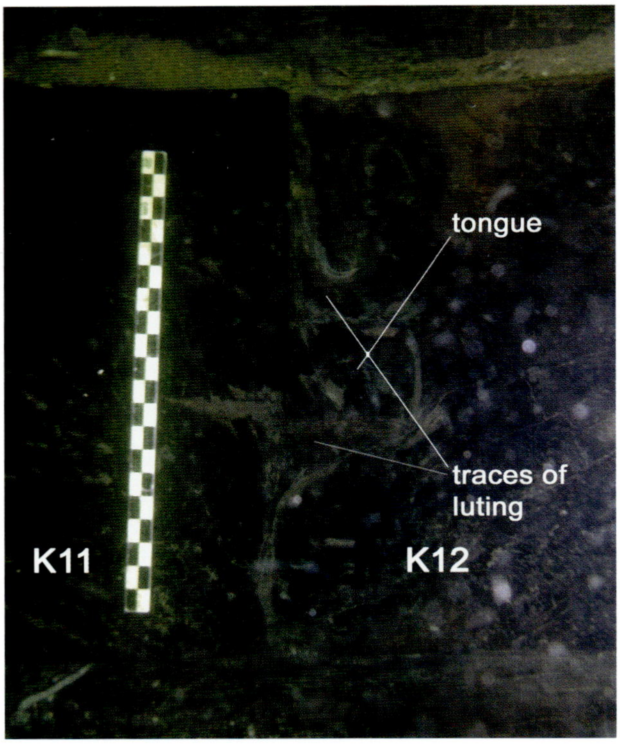

Figure 4.10 Jont between segments K11 and K12 viewed from above. (Photo: A. Belov; © F. Goddio/IEASM.)

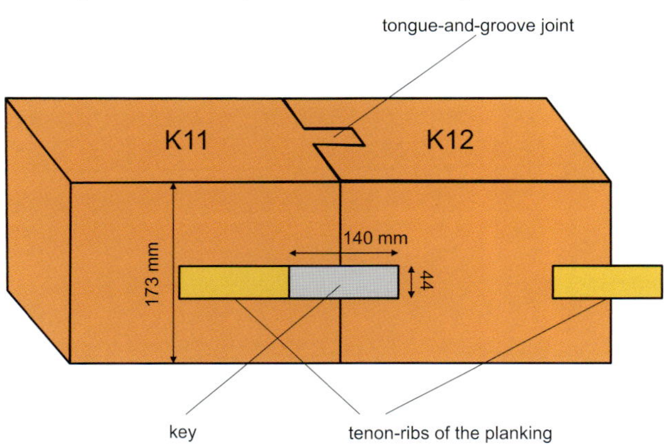

Figure 4.11 Joint between segments K11 and K12 reinforced by a lateral key. (Drawing: A. Belov; © F. Goddio/IEASM.)

keels can be seen in some of the models of travelling ships from the tomb of Tutankhamun (1336–1327 BC).[20]

Jones remarks that the presence of 'elongated finials' does not prove the existence of a keel and cites the example of the Khufu I ship where the finials are fastened to short planks at the extremities called 'backing timbers'.[21] On these grounds, Jones denies that Egyptian ships of the eighteenth dynasty included a keel in their construction. This argument is inconclusive, however, as data from a (probably unfinished) papyriform royal barque[22] can hardly be extrapolated to travelling ships and warships. In addition, the sheer view of the Khufu I ship is quite different to that of the above-cited models that show two distinct curves: one corresponding to the line of the keel and the other to the line of the planking.

Bradbury[23] concludes that 'These models [two boat models found in the tomb of Amenhotep II] imply that they [the Egyptians] had incorporated the single timber keel in some of their travelling ships, and certainly those sailing on a high sea. It was about this time too they adopted the Levantine seagoing *mnš*-ships which did possess a keel'. This agrees with Landström's suggestion that the concept of a keel was brought to Egypt by Palestinian or Syrian shipbuilders.[24] Syrians were employed in the shipyard and at the main Egyptian port of *prw-nfr* in Memphis in the reign of Thutmose III (1479–1425 BC)[25] and, probably, Amenhotep II (1427–1400 BC) as well. The gods Baal and Astarte were venerated at *prw-nfr* indicating that the Syro-Canaanites were firmly installed in this area.[26] While it is debatable whether the Egyptian term *pipit* referred to a keel,[27] it is more certain for *tp-ḫ.t*, which is attested for the first time in the wood register of the *prw-nfr* shipyard.[28] One can hardly fail to see the parallel with the term *tpy-ḫ.t* found in the 'Story of Wenamun',[29] which was probably composed during the twenty-second dynasty (945–*c.* 715 BC), although it was set in the reign of the twentieth-dynasty pharaoh Ramesses XI (*c.* 1099–1069 BC). Wenamun, a priest of Amun at Karnak, was sent to Lebanon to buy the cedar necessary to build the divine barque of Amun Userhet. The text alludes to the construction of an earlier barque in the time of Ramesses III (1184–1153 BC) and mentions that the keel of this ship measured 130 cubits in length (about 68 m).[30] Bradbury judiciously remarks that the keel of this river-going vessel must have been composed of several segments scarfed together.[31]

Iconographic documents have allowed a date to be proposed for the appearance of *menesh-* (*mnš*) ships in Egypt.[32] The construction of Syrian ships depicted in the tomb of Nebamun (reign of Amenhotep II) and Kenamun (reign of Amenhotep III) has been extensively discussed in the literature.[33] Among some typically Egyptian features, these ships also show obvious elements of foreign origin, such as the vertical endposts of the hull, the fence-like gunwale, a foreign rigging arrangement, and a large Canaanite amphora for storing drinking water. The absence of a hogging truss is contrary to the construction

20 Landström 1970: 107, figs 331–3.
21 Jones 1995: 54.
22 See Mark 2009: 148–50; 2011: 27.
23 Bradbury 1996: 50.
24 Landström 1970: 107.
25 *P.Ermitage* 1116 B, verso 15.
26 *P.British Museum* 10056: column 4, line 3; column 9, line 12, commented in Glanville 1932: 25–6, 29; see also Säve-Söderbergh 1946: 37.
27 Jones 1988: 164.
28 *P.British Museum* 10056, column 6, line 12; see Bradbury 1996: 51 note 69.
29 Glanville 1932: 27 note 82; Goedicke 1975: 142–3 note 156.
30 Goedicke 1975: 142–3 note 156.
31 Bradbury 1996: 51 note 69.
32 Basch 1978: 115–8.
33 Ibid. 1978; Bradbury 1996; Davies and Faulkner 1947; Edgerton 1931; Köster 1923; Landström 1970: 139; Säve-Söderbergh 1946: 56–7.

5 Planking

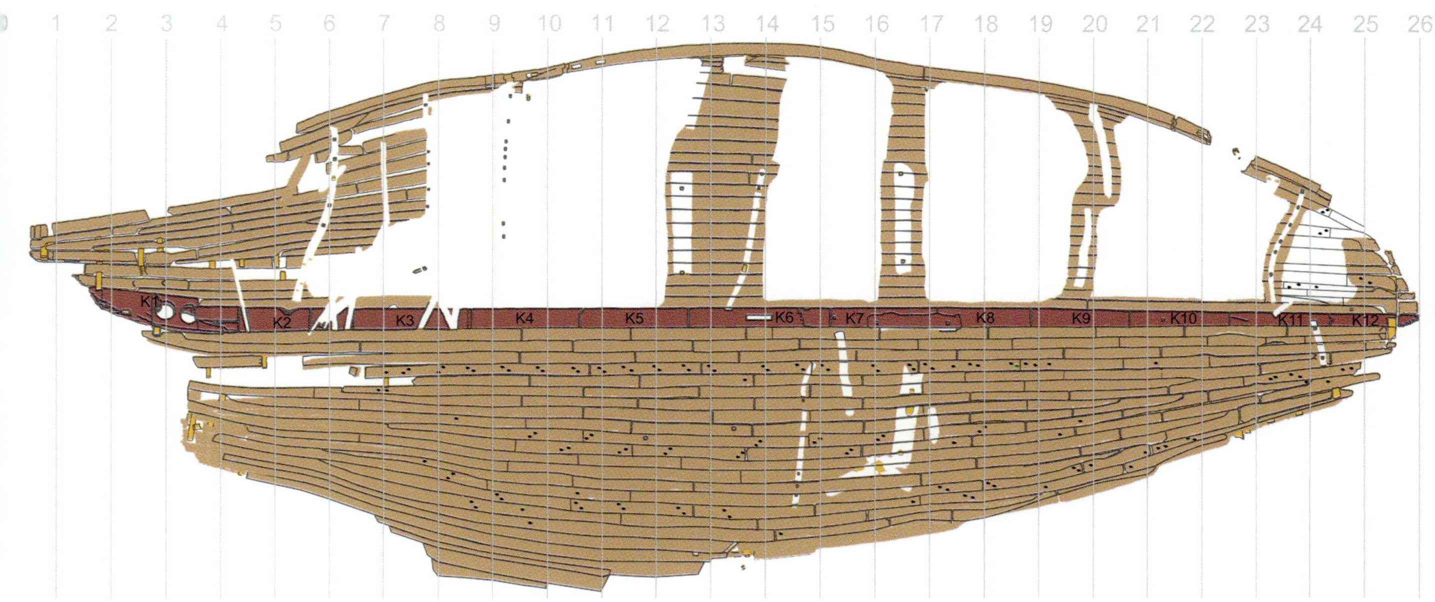

Figure 5.1 Planking plan of ship 17 from Thonis-Heracleion. (Drawing: P. Sandrin/A. Belov; © F. Goddio/IEASM.)

Plan of the planks

Twenty-four lines of strakes were preserved on the starboard of ship 17 and twenty-two on the port (Figure 5.1). The latter was not excavated completely and so the data on the planking was mainly collected on the starboard. From the excavated sections of the port side, however, it is clear that the arrangement of the planking was not symmetrical.[1]

The most remarkable feature of the planking was the phenomenon called 'imbrication' by Boreux.[2] This method provided a regular alternation between the joints (staggering of the joints) of the neighbouring strakes in order to give maximum solidity to a hull built with short planks (Figure 5.2). The plan of the planks is characterised by many stealer strakes, some of which ended with knife-shaped planks (Figures 5.3–5.4). A single plank was shaped to curve around the end of the adjacent plank thus reducing two strakes to one. There were eight strakes of this type on the starboard: S4, S6, S8, S9, S10, S12, S13, and S14. They serve as a good illustration of a common boatbuilding principle of diminishing the number of strakes towards the extremities of a vessel.

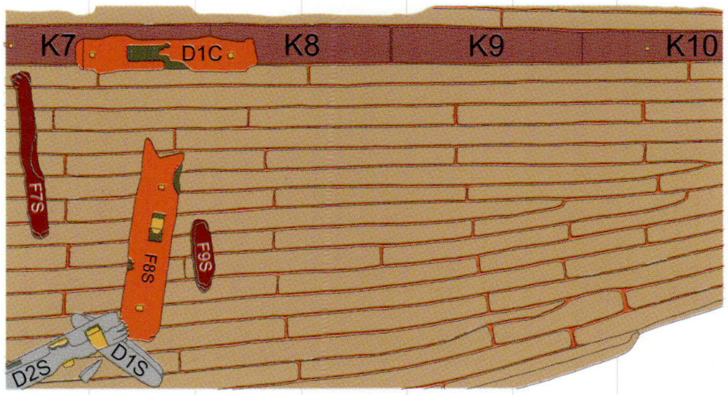

Figure 5.2 Fragment of the hull showing imbrication. (Drawing: P. Sandrin/A. Belov; © F. Goddio/IEASM.)

The form of the planks

There are four different types of planks observable that in combination make up the hull of ship 17. Type A planks are rectangular shape in plan and were by far the most common form. Type B are 'knife-shaped' and make up the last bow plates of stealer strakes S9, S11 and the last stern stealer strakes S7, S10, and P10. Type C planks taper towards the extremity of the hull. Type D planks

1 Without reopening here a discussion on 'shell-first' and 'frame-first' types of construction (first proposed by Hasslöf 1963), and on the transition between them (recent publications: Olaberria 2013; 2014; Pomey, Kahanov and Rieth 2012; 2013; McGrail 2013), the asymmetry of the hulls of ancient ships may be illustrated by the fourth-century Kyrenia wreck and the first-century Madrague de Giens wreck (see Pomey 2004: 50).

2 Boreux 1925: 248.

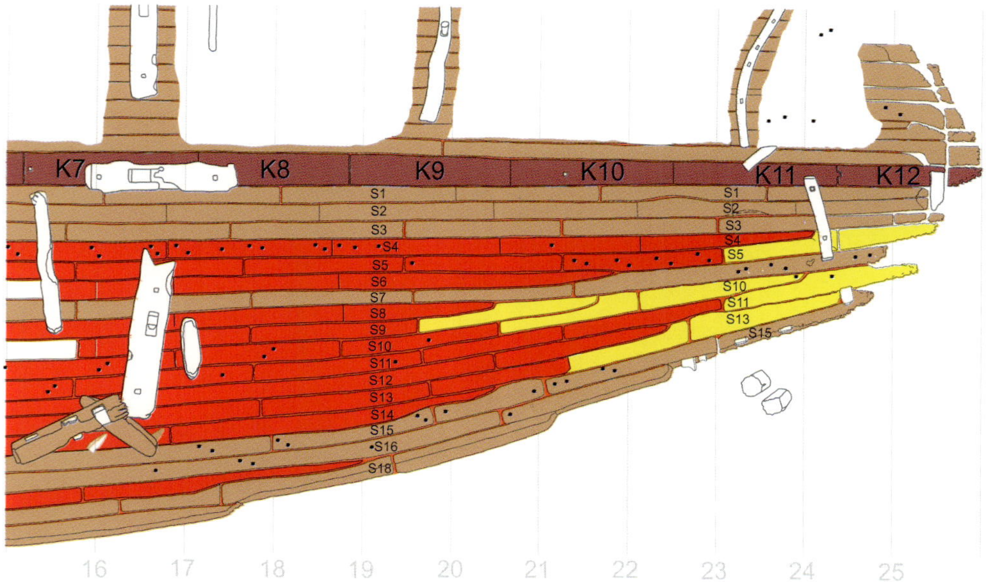

Figure 5.3 Stealer strakes (in red) ending with knife-shaped planks (in yellow). Starboard bow part of the hull. (Drawing: P. Sandrin/A. Belov; © F. Goddio/IEASM.)

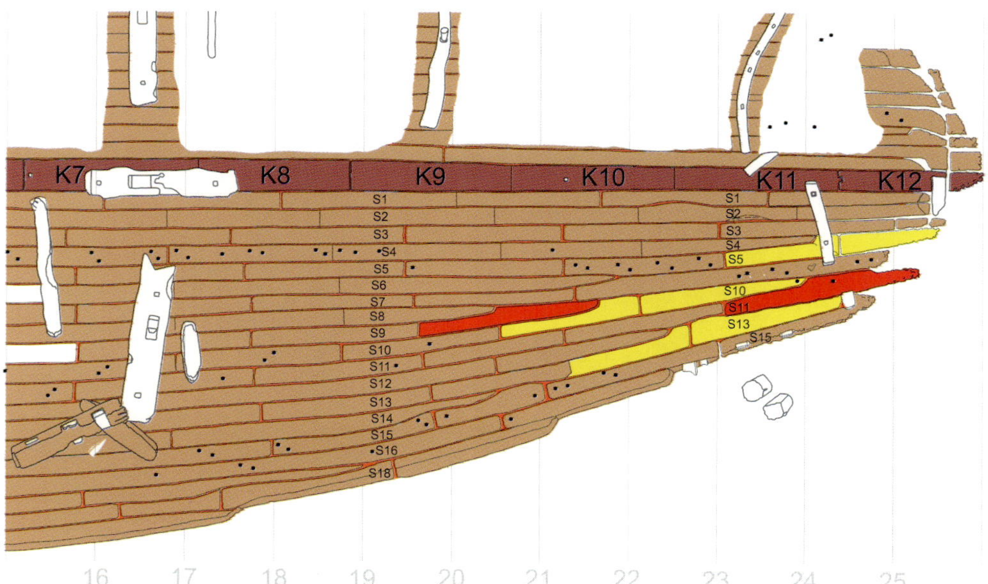

Figure 5.4 Planks of types 'C' and 'D' (in yellow, left-C, right-D) and knife-shaped planks of 'B'-type (in red). (Drawing: P. Sandrin/A. Belov; © F. Goddio/IEASM.)

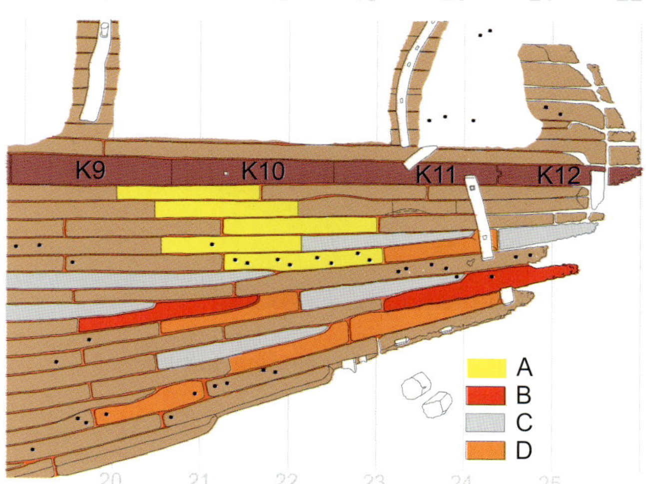

Figure 5.5 Planks of type 'C' (in grey). (Drawing: P. Sandrin/A. Belov; © F. Goddio/IEASM.)

are 'club-shaped'. All of the different planking types are illustrated in Figure 5.5. This form of planking was used in many other examples of ancient Egyptian boats.[3]

3 Pomey 2012a: 39–40; 2012b: 10; Ward 2000: 117, fig. 36.

The inboard surfaces of the planks appeared to be flat (Figure 5.6), while their outboard surface was sometimes slightly curved. Many planks followed the grain of the wood and contained knots and fissures.

The dimensions of the planks

The length of the planks was measured on the starboard side of ship 17. In some cases it was not possible to determine the position of the joint between the planks that were hidden by elements of internal structure. Consequently, uncertain data was not included in the calculations below that were based upon measurements of 102 starboard planks, which corresponds to approximately 75 per cent of its preserved surface. The width of planks was measured at 1 m intervals along the keel and included planks in the excavated regions of the port side of the side. Unfortunately, the data on the thickness

Figure 5.6 View of the inner hull at the starboard (13–17 m along the baseline). (Photo: C. Gerigk; © F. Goddio/IEASM.)

of planks is rather poor and measurements were taken from the upper preserved strakes, those exposed in disintegrated parts of the hull and at several sections.

Length

The length of the planks on the starboard side varied from 0.49 to 3.77 m (Table 5.1). The majority of planks, 75 per cent of all starboard planks, measured between 1.70 and 2.20 m in length. The average length of the starboard planks was 192 cm.

It is possible to suggest several dimensional groups according to the length of the planks (Table 5.2). The first group includes all planks shorter than 1.70 m, which seem to have been used as inserts between longer planks. The largest group (75 per cent of all planks) was formed

Table 5.1 Length of planks of the starboard of ship 17 (cm). Colour corresponds to the dimensional groups discussed in the text.

	Bow								Aft	
Plank No.	1	2	3	4	5	6	7	8	9	10	Average strake
Strake											
S1	189.4	183.9	160.1	195.2	193.5		49	207.4	196.2		171.84
S2	165	70	110.2	180.1	186	185					149.38
S3	194.5	171.8	173.8	191.8	181.8	211.8	186.2	208.1			189.98
S4	194.1	157.1	187.4	182.6	191.2	199.8	201.1				187.61
S5	126.7	177.9	175.6	177.4	186.3	195.6	205.1	197.8	212.2		183.84
S6	307.5	184.7	202.8	200	208.8						220.76
S7	215.4	135.8	177.9	179.2	193.4	195.5	209.5	209.6	377.3	228.8	212.24
S8	167.6	189.2	206.3	176.6	233.5		244.7				202.98
S9	201.3	185.4	191.2	200.4	217.3	190.7					197.72
S10	186.1	157.3	175.6	183.9	208.7						182.32
S11	220.6	179	157.6	187.6	199.1	207.6	205.2	212.8			196.19
S12	195.1		187.6		201.2						194.63
S13	172.4	146	156.9	188.1							165.85
S14			219.2								219.20
S15	180.1	196.2	125.5	181.4		261.9					189.02
S16	194		192.1	219.3	222.5						206.98
S17	313.7	198	248.1								253.27
S18	182.9	194.4	112.1	206.7							174.03
									Min	Max	**Average starboard**
									49	377.30	191.82

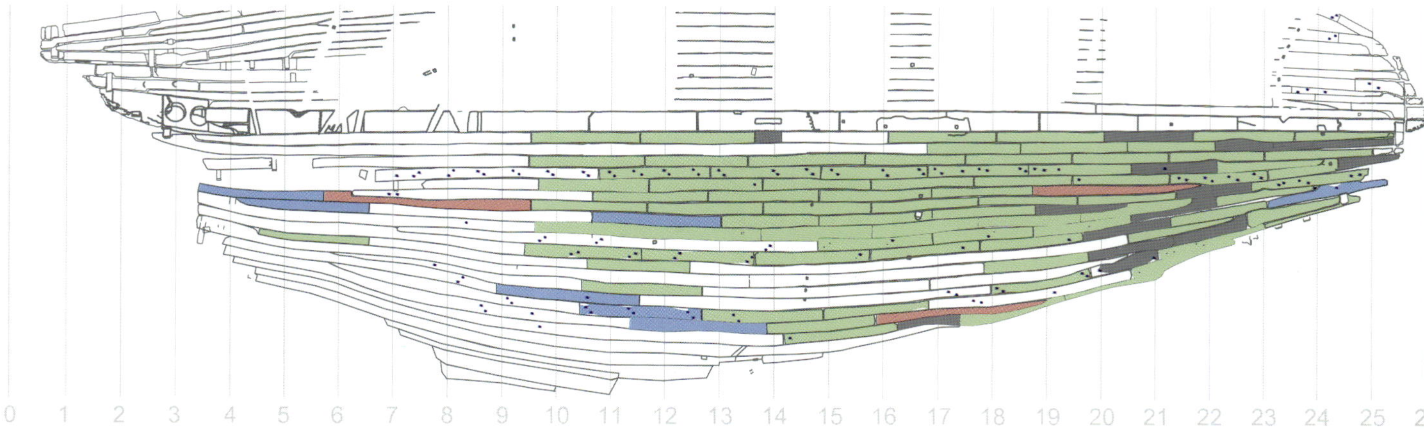

Figure 5.7 Dimensional groups according to the length of the planks. The colour of the planks corresponds to those of the Table 5.2. (Drawing: A. Belov; © F. Goddio/IEASM.)

Table 5.2 Dimensional groups according to plank length (cm) of the starboard.

Group	Number	Percent	Avg. length (cm)
<170	15	14.7	133.1
170–220	77	75.5	193.6
220–270	7	6.9	237.2
>270	3	2.9	332.8

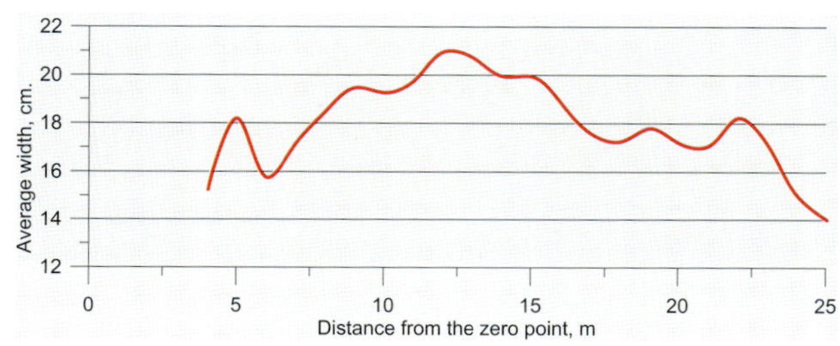

Figure 5.8 Average width of planks along the keel.

by the planks with a length between 1.70 and 2.20 m with an average value of 1.94 m. Planks longer than 2.20 m were few in number (10 planks, 10 per cent) and were characteristic of the extremities of the ship (Figure 5.7).

Width

The width of the planks on the starboard side of ship 17 varied between 9 and 29 cm with an average value of 19 cm (Table 5.3). On the port side, the plank width varied between 9 and 27 cm, with an average value of 18 cm (Table 5.4). In both instances the largest planks were found closer to the keel in the centre of the hull (Figure 5.8).

Thickness

The thickness of planks of the upper preserved strakes ranged from 12 to 18 cm with an average value of 14 cm (Table 5.5). Approximately the same figures on thickness were obtained during the transversal cuts of the hull (Table 5.6). It is to be noted that the same plank may have had changing thickness within the limits of 2–3 cm.

Discussion

The method of 'imbrication', the staggering of the joints between planks, allowed the Egyptians to build a solid hull using only very short planks[4] and was described by Herodotus (*Historia* 2.96) in relation to construction of a *baris*.[5] There also exists numerous representations of 'brick-like' planking patterns in ancient Egyptian art. The first of them are seen during the twelfth dynasty (1985–1773 BC), although it is probable that the method of imbrication was already in use in the fourth to sixth dynasties (2613–2181 BC).[6] The latest known representations of imbrication are dated to the end of the New Kingdom (Figure 5.9) and thus the evidence of ship 17 confirms that this process continued in the fifth-fourth centuries BC. It would seem that the lack of good, long lengths of local timber lies behind the development of the 'imbrication' method in ancient Egyptian shipbuilding.

Figure 5.9 A boat from the tomb of Ipy, probably twentieth dynasty (c. 1186–1069 BC). (Drawing: A. Belov; after de Garis Davies 1927: pl. 30.)

4 Vinson puts it in the following terms: 'To avoid areas of weakness, the joints of the individual planks making up the strakes were placed so they did not line up vertically with joints of the strakes above and below. This staggering of the joints is what creates the "brick-like" impression of the planking pattern' (1998: 255).

5 See this volume, Chapter 10.

6 Boreux 1925: 248–9.

Table 5.3 Width of planks of the starboard (cm). Position from the zero point.

Position / Strake	26	25	24	23	22	21	20	19	18	17	16	15	14	13	12	11	10	9	8	7	6	5	4	Average (strake)
S24																	[9]	[12]	25					
S23																27	29	26	?					27.3
S22																14	10	?	?					12.0
S21														23	28	13	13	?	?					19.3
S20												[11]	[14]	[9]	22	23.5	?	?	?	18				21.2
S19										[7]	[13]	21	16.5	25	20	22	?	?	?	15				19.9
S18						[11]	[5]	[5.5]	[14]	13.5	17	16.5	22	23	24	25	21	?	?	9.5				19.1
S17									13.5	14	20	22	20.5	17	23.5	17	22	22	?	?	19			19.1
S16					[11]	19	18.5	18	[21]	19	15	?	20.5	22	22	22	20	19	?	25.5	[7]			20.0
S15			[5]	[10]	10	22.5	18	?	?	14	?	?	22.5	24	20	21	19.5	19.5	?	18	14.5			18.6
S14							16	20	?	?	?	?	18	16.5	20	18.5	22	19.5	13.5	10	19			17.5
S13			15	24	29	18	16	17	?	?	18	?	23	22.5	21.5	23	17	19	17	16	17.5	[7.5]		19.6
S12					9	16.5	20.5	18	?	?	17	20	15	21	20	17	16	18	18.5	16.5	14	13.5		16.9
S11		[12]	19	13	20	18	15	17	19	17	18.5	18	24	20.5	20	18.5	21.5	20	18.5	22	15	24		18.9
S10			[9]	17	22	11	18	17	22	23.5	21	24.5	20	?	20.5	21	22	18	13.5					19.4
S9						18	11.5	18	16	21.5	23	20	21.5	?	21	21	17.5	20	18	21.5	16	18		18.9
S8							13.5	18.5	15.5	14	15.5	19.5	17	21	19	19	22.5	18	15.5	20	16	20		17.8
S7		20	20	13	20	13.5	17.5	15	15	17	18.3	21	20.5	17	17.5	17	18	22.5	17	11.5	12.5	16		17.1
S6						14	16.5	18.5	14.5	18	19	22	16	17	18	19	17.5	16.5	20	13				17.3
S5		14	16	20	22	18.5	17	17.5	18.5	16	20	16	21.5	21	25	20	19.5	20						19.0
S4			10	14	19	17	20	17	18	21	23	23	21	21	19	20	18.5	15	20					18.6
S3		13	16	17	14	18.5	19	16.5	19	17	17.5	18.5	18	18	19	17	18	21.5	21	18.5		18		17.8
S2		12	13	16	17	16.5	19.5	20	19	20.5	16	17.5	21.5	21.5	19.5	21	20	16	20	20	18	16	18	18.1
S1		11	17	20	18.5	18	18.5	19	17	17.5	22	19.5	20	22	21	18.5	20	20	21.5	20	20.5	20	12.5	18.8
Keel	19	22	25	28.5	29.5	33.5	35	35	37.5	37.5	35.9	37.7	38	37	36	36	37.5	38.5	36	37	41	43.5	53	35.2
Average (section)		14.0	15.8	17.1	18.2	17.1	17.2	17.8	17.3	17.6	18.8	19.9	19.9	20.7	21.0	19.8	19.3	19.5	18.5	17.2	16.5	18.2	15.3	

[10] – preserved width ? – width unknown [shaded] – outline of excavated planking

Table 5.4 Width of planks of the port (cm). Position from the zero point.

Keel	19	22	25	28.5	29.5	33.5	35	35	37.5	37.5	35.9	37.7	38	37	36	36	37.5	38.5	36	37	41	43.5	53	53	[32.1]	
Strake No.																										
1			15.5				17			17.5			15							20.5		14			19.5	
2			17				16			16.5			18							?		14			20.5	
3			19.5				21			18			16							?		17			17	
4			18.5				18			19			21							?		?			9.5	
5			[6]				21.5			18			16							?		20			11	
6			21				15.5			21			20							14		17			12	
7			20				16.5			19			18							13.5		15.5			21.5	
8			15.5				17			16.5			20							12.5		16				
9			18.5				18			16.5			17							12		15				
10			16.5				17.5			19			21							12.5		13.5				
11			19				19.5			21.5			19							13.5		14				
12			17				19.5			23			21							17		15				
13			16				12			18			17							20.5						
14			[3]				14.5			21.5			23							21.5						
15							19			21.5			20							21						
16							24			19.5			16							19						
17							26			19			27.5							?						
18							21			18.5			21							13						
19							9.5			17.5			21.5							1.5						
20							9			11			17							[6.5]						
21							[4.7]			13			22													
22										[4.5]			22													
23													[14]													
24																										
Position	26	25	24.5	23	22	21	20	19	18	17	16	15	14	13	12	11	10	9	8	7	6	5	4	3	2	1

[10] – preserved width ? – width unknown [shaded] – outline of excavated planking

Table 5.5 Thickness of planks of upper preserved starboard strakes (cm). Position from the zero point.

Position	26	25	24	23	22	21	20	19	18	17	16	15	14	13	12	11	10	9	8	7	6	5	4	3
Strake		15	15	16	18		18	18		19	20	20	19	20	20	23	23	22	23	23	20	19	16	3
Thickness, cm	?	12	12	14	14	?	14	12	?	15	17	16	18	14	14	15	12	12	16	15	14	12	14	14

Table 5.6 Thickness of planks measured at the section in the bow starboard area of the ship (21 m along the baseline) (cm).

Strake	16	15	13	12	11	10	9	7	6	5	4
Thickness, cm	17.5	14.5	15.3	13.9	15.7	12.7	14.5	13.9	14.3	15.5	13.9

Table 5.7 Main dimensions of the knife-shaped planks from Lisht and Mersa/Wadi Gawasis and from ship 17.

Planks	Length, cm	Width (max), cm	Thickness, cm
Licht, type I (e.g. timber 3)	200–268	35–39.5	15–18
Mersa Gawasis, T34	293	46	15
Ship 17, type B	202–221	19–24.5	15–18

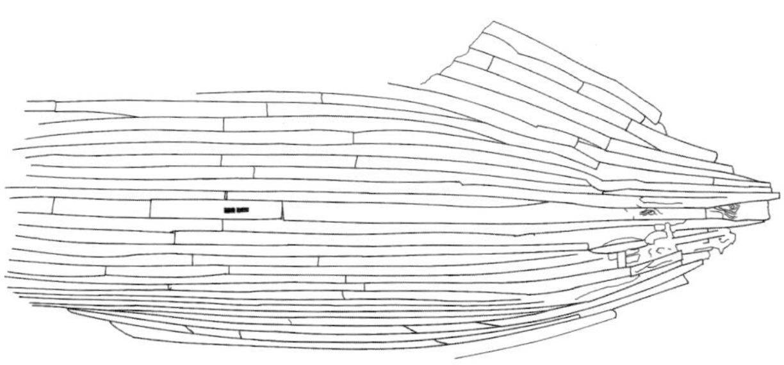

Figure 5.10 Sketch of the hull of the Mataria boat (fifth century BC). (Drawing: C. Ward 2000: 130. Courtesy C. Ward.)

Indeed, local species like acacia, sycomore fig, tamarisk and others did not allow long straight planks to be cut from them and imported wood was expensive.[7]

The planking plan of ship 17 had many common features with that of the Mataria boat, discovered during construction works in Heliopolis.[8] The boat was dated to the middle of the fifth century BC, which makes it almost contemporary with ship 17. Its planking, cut from local sycomore fig, also showed numerous stealer strakes and a staggering of the joints (Figure 5.10).

The form of knife-shaped planks of ship 17 (type B) had much in common with type I planks in a freighter from Lisht constructed of tamarisk and acacia and dated to the twelfth dynasty (1950 BC).[9] Plank T34 from Mersa/wadi Gawasis, belonging to a seagoing ship built of cedar (*Cedrus libani*), was much more angular.[10] The length and thickness of all above-mentioned planks were quite close to those of ship 17 (Table 5.7), although the type B planks of ship 17 were not as wide.

Ward suggested from fragmentary data that planks from the near-contemporary Mataria ship were from 0.80 to 6.0 m in length, with the majority being between two and four metres long.[11] The planks were about 20 cm wide.[12] Thus, while the width of the planks of the Mataria ship are very similar to ship 17, their length on average was greater, with the biggest planks being roughly three times longer than those used in the ship from Thonis-Heracleion. In the Mataria boat, being a smaller craft than ship 17, this difference may be attributed to the longer timber that can be more easily obtained from sycomore fig.[13]

According to Herodotus, the length of a plank of a *baris* was two cubits, i.e., around 104 cm.[14] For ship 17, however, the planks were almost twice as long.[15] Boreux was the first to point out that much longer planks can be obtained from the acacia tree and that Herodotus was likely describing relatively small specimens of the *baris*.[16] According to Theophrastus, planks up to 12 cubits long (about 6.24 m) can be obtained from the acacia tree.[17] The reconstructed length of ship 17 is around 28 m and it obviously belonged to a class of larger cargo boats, and it seems logical to suppose that longer planks of acacia were necessary to build a larger hull. However, it must be remembered that the planks of ship 17 were still very short compared to the standards of Greco-Roman shipbuilding elsewhere in the Mediterranean.[18] Thus

7 Ward 2000: 24.
8 Haldane 1996b: 441; Ward 2000: 129–35.
9 Ward 2000: 117.
10 Fragments of ships from Mersa/wadi Gawasis are not easily dated because of continuous occupation of the site from the Old to the New Kingdom. Thus ship timbers can belong to the Middle or New Kingdom. See Ward and Zazzaro 2010.
11 Ward 2000: 131.
12 Ibid.
13 It is notable that the large obelisk barges of Queen Hatshepsut were also built of sycomore fig.
14 It would be difficult to imagine a hull completely built of planks of a same length; however, it is quite admissible that two-cubit planks could make up the majority of a hull: see Rougé 1975: 41. An ethnographic parallel can be cited. The planks of the boats of the Upper Nile built of *Acacia nilotica* and called *nuggar* (*naggr*) are usually 4–6 feet (1.2–1.8 m) long and never exceed 8 feet (2.4 m) in length. Hornell 1943a: 28: 'The rougher and cheaper the boat, the shorter are these planks'.
15 The average length of planks of the main dimensional group (170–220 cm) makes 193.6 cm.
16 Boreux 1925: 237. Indeed, lines 22–23 of Herodotus *Historia* 2.96 suggest the existence of *baris*-ships of quite variable tonnage.
17 Theophrastus *In Historia Plantarum* 2.4.2.8.
18 In Mediterranean boatbuilding traditions, the choice of wood species was very wide. Coniferous species were preferred for the planking because of their flexibility (Guibal and Pomey 2002: 96). Long planks were obtained from pine, fir, cedar, cypress and larch (Theophrastus *Historia Plantarum* 4.1.2–4, 5.1.5–8, 5.8.1–3. Other references to the ancient sources can be found in Rival 1991: 90, Index). In normal shipbuilding practice, the longer the planks are, the better. The length of the majority of completely preserved planks of pine of the Kyrenia ship (295–285 BC, reconstructed length 14 m), varies from 8 to 10 m: see Steffy 1985: 79, table 1.

Figure 5.11 Top view of upper preserved strake P23 showing three channels. (Photo: C. Gerigk; © F. Goddio/IEASM.)

the planks from ship 17 still sit within a local, purely Egyptian, tradition of assembling the hulls with short timbers[19] that were at the same time rather wide[20] and very thick. The width of the planks of ship 17 varies from 9 to 29 cm with an average value of 18.2 cm. Thus the planks were characterised by a ratio between their width and length of about 1:10, which increased the effect of the brick-like appearance of the planking.

The assembly of the planks

The most intriguing part of the construction of ship 17 was the manner in which the planking was assembled. The planks were joined transversally using very long tenons. They were installed in rectangular channels cut into the centre of the plank edges and passed through several strakes.

Channels

Rectangular channels[21] were cut with a chisel into the centre of the plank from opposite sides, meeting in the middle (Figure 5.11). The exceptions to this are the channels that pass through the keel, which were cut into the lower third of the segments. The length of the channels ranged from 11 and 31 cm with an average value of 20 cm. The width changed between 4 and 8 cm (6 cm on average). The distance between the centres of channels varied from 27 to 70 cm (47 cm on average). Two types of channels are easily distinguished: those that were destined to receive only one tenon and longer ones that received two lateral tenons. The average length of the former type was 19 cm, the latter 31 cm.

Tenons

While the majority of tenons were rectangular in section, some of them were trapezoidal or oval. Like the planking, they were also cut from acacia. In many cases tenons were wedged inside the channels in some of the strakes with small slips of wood, triangular in profile and rectangular in section (Figures 5.12a-b). The length of wedges varied from 15 to 17 cm, their width from 1 to 6 cm and thickness from 2.5 to 6 cm.

Wedges were also found with the tenons passing through the keel. In this case there were always four wedges placed symmetrically from the both sides of each tenon (Figures 5.13–5.15). It seems that this procedure allowed the maintenance of the immobile and precise position of these primary tenons inside their channels,

19 It seems that in some cases Egyptian shipbuilding continued to use short planks even when the type of wood allowed for longer timbers: see Vinson 1998: 255.

20 Ibid.: 255 note 19.

21 Following the suggestion of C. Ward, this term was preferred to 'mortise'. A mortise is determined as a cavity cut into a timber to receive a tenon. There exist 'through-mortises' that pass entirely through a piece. However, 'channel', better conveys the fact that the openings were aligned in numerous strakes to receive one long tenon.

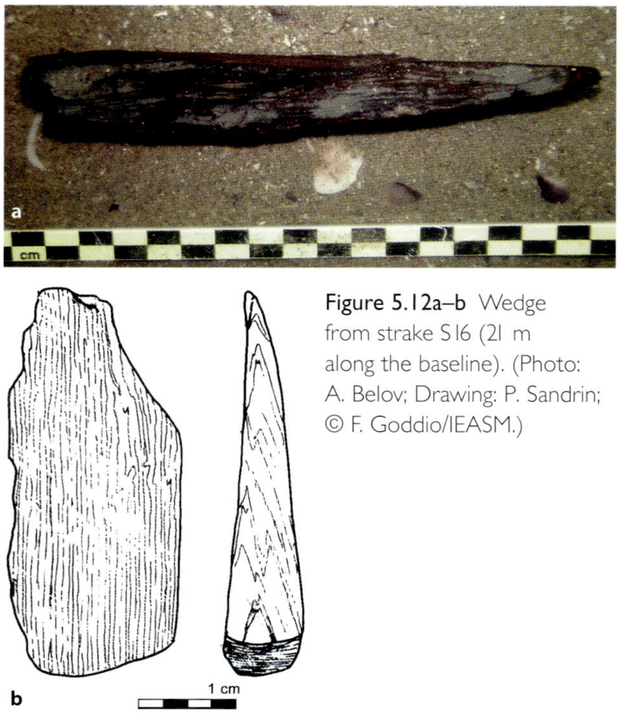

Figure 5.12a–b Wedge from strake S16 (21 m along the baseline). (Photo: A. Belov; Drawing: P. Sandrin; © F. Goddio/IEASM.)

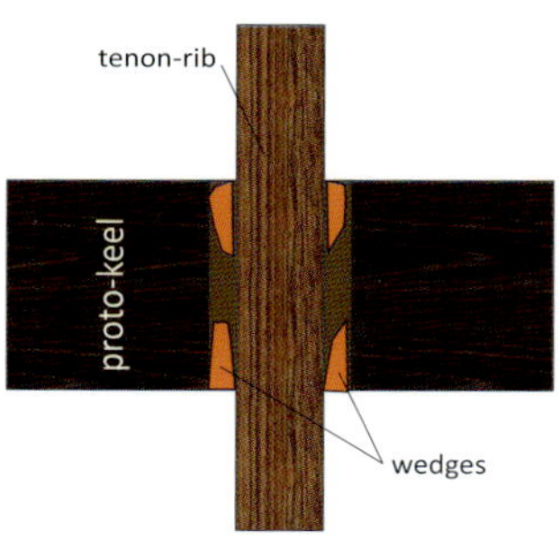

Figure 5.13 Scheme showing the interior of one of the channels of the proto-keel (viewed from above). A tenon was secured within the channel with four wedges. (Drawing: A. Belov.)

Figure 5.14 Fragment of tenon of the planking (right) and the wedge (left) that secured it inside the channel of segment K6 of the proto-keel of Ship 17. Pieces were removed from the channel for clarity. (Photo: A. Belov; © F. Goddio/IEASM.)

Figure 5.15 Side view of segment K6 of the keel with a tenon wedged from both sides inside its channel. (K6 is shown upside down). (Photo: A. Belov; © F. Goddio/IEASM.)

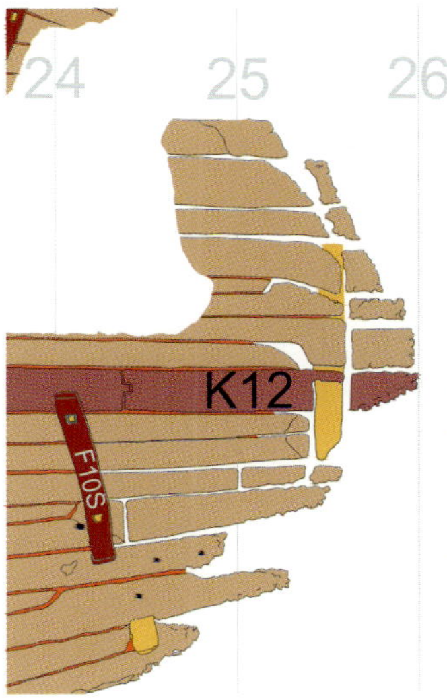

Figure 5.16 Plan of tenon of the planking at the preserved end of the bow, marked in yellow (25.2 m along the baseline). (Drawing: P. Sandrin; © F. Goddio/ IEASM.)

which was important for the subsequent building of the shell and the rigidity of the vessel.

The considerable length of the tenons became evident from the beginning of excavation. A partially preserved tenon 108 cm long was documented at the bow (Figure 5.16–5.17) and another one, measuring 80 cm, at the stern (Figure 5.18). These tenons were apparently not preserved to their full length. In order to document a complete tenon, one was chosen on the starboard side, 9 m from the preserved bow extremity (Figure 5.19). A section was made at strake S16 and the channels were half-opened on the inboard side of the hull using a saw. Double-size channels were discovered in strakes S16 and S15. They contained the remains of a tenon that continued in the outboard direction and the extremity of

Figure 5.17 Top view of tenon of the planking at the preserved end of the bow (25.2 m along the baseline). (Photo: A. Belov; © F. Goddio/ IEASM.)

Figure 5.18 Tenon of the planking at the preserved end of the stern (1.8 m along the baseline). (Photo: C. Gerigk; © F. Goddio/IEASM.)

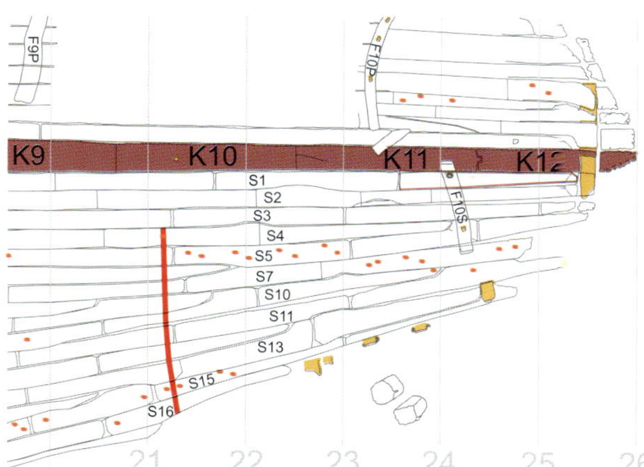

Figure 5.19 Position of a transversal cut made on the starboard at 21.3 m along the baseline (red line). (Drawing: P. Sandrin/A. Belov; © F. Goddio/IEASM.)

another tenon that continued in an opposite direction, i.e. towards the keel. Thus, these tenons were superimposed laterally in two strakes, S15 and S16 (Figure 5.20).

Figure 5.20 Top view of two superimposed tenons discovered in strakes S15 and S16 (21.3 along the baseline, on the starboard). (Photo: A. Belov; © F. Goddio/IEASM.)

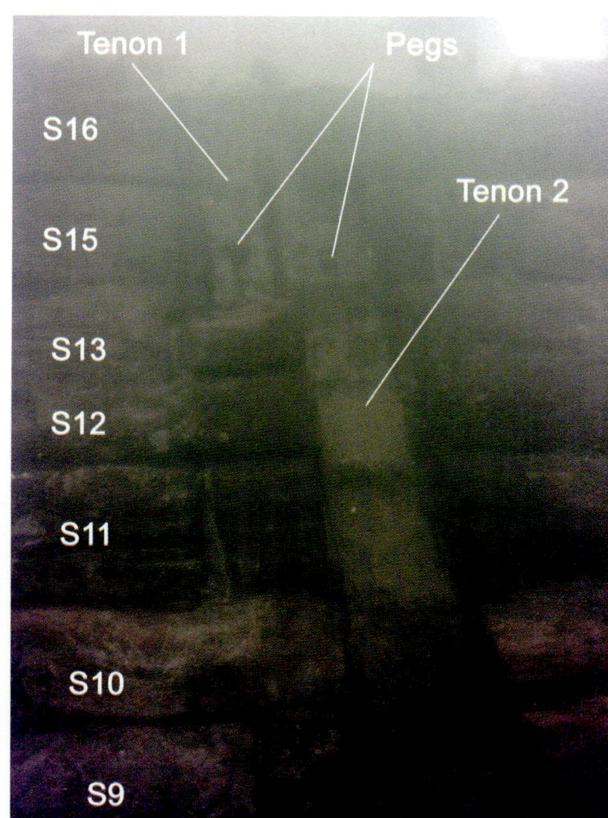

Figure 5.21 Top view of two superimposed tenons discovered in strakes S16–S4 (21.3 along the baseline, on the starboard). (Photo: A. Belov; © F. Goddio/IEASM.)

Figure 5.22 Top view of the ends of the tenon pegged in strake S4. (Photo: A. Belov; © F. Goddio/IEASM.)

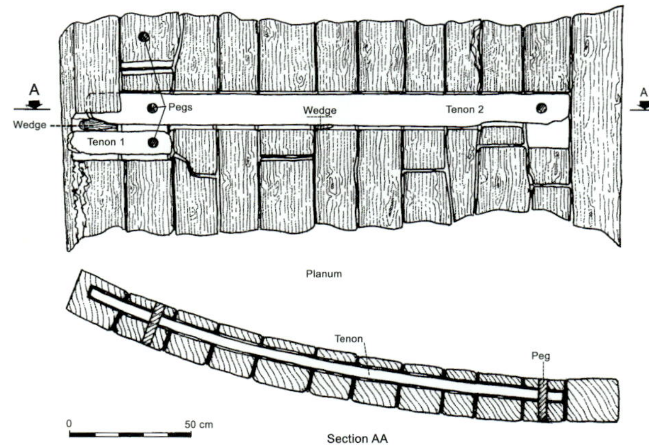

Figure 5.23 Plan and profile of the section at the bow. The strakes were opened from the interior of the hull to reveal the tenon *in situ*. (Drawing: P. Sandrin; © F. Goddio/IEASM.)

A second tenon, preserved in its entirety, was followed towards the keel to its end in strake S4. Like the tenon previously examined, this was also attached to the planking with round pegs 35 mm in diameter (Figures 5.21–5.22). The pegs went through the planking and were visible from the outside of the hull. The S4 to S16 tenon measured 199 cm in length and passed through 11 strakes of planking (Figure 5.23). The width of the tenon varied between 11.5 and 13 cm, with a thickness of 3.5 cm. It was noted that the tenon was not always in the middle of a channel within different strakes, but had a slight sideward shift. In my opinion, this provides evidence that the tenon was first installed in the keel and

Figure 5.24 Fracture in the tenon between strakes S5 and S6 (white arrows). (Photo: A. Belov; © F. Goddio/IEASM.)

Table 5.8 Width and thickness of tenons of the planking and of wedges within the segments of the keel (cm). (Distance along the baseline. Numbering of channels starts from the stern).

	Keel segment K6					S1, 13m	S1, 14–16m		S1, 24–25m				Average cm
Mortise no.	1	2	3	4	5	1	1	2	1	2	3	4	
Width, cm	12.0	12.2	11.1	13.0	12.1	11.6	8.4	12.0	13.0	12.5	14.5	15.5	12.3
Thickness, cm	4.9	4.2	5.0	3.0	5.2	4.4	5.0	5.0	4.0	4.9	4.8	4.0	4.5
Wedge 1, width	1.2	?	2.1		1.3	3.3	1.6		?		?	?	1.9
Wedge 1, thickness	5.1	?	5.0		5.2	4.4	5.0		?		?	?	4.9
Wedge 2, width	1.2	1.9	2.1		1.3	1.5	4.4	5.0					2.5
Wedge 2, thickness	5.9	4.2	5.0		5.2	4.4	5.0	5.0					5.0

then the channels were cut in the planks of each strake that was to be added next. The tenon was wedged within the channels of strakes S16, S12 and S10. The wedge in strake S16 was placed between two neighbouring tenons. The ends of the tenon were roughly rounded (see Figure 5.22). The tenon had visible fractures in line with the strakes from S4 to S6 that were likely related to the post-depositional deformation of the hull (Figure 5.24).

Data on the width and thickness of tenons and on the dimensions of the wedges are presented in Table 5.8. The figures derived from the study of the tenons associated with the keel are similar to those measured in the channels of the strakes (Table 5.9). The measurements of tenons found in the upper preserved strakes are less reliable because of their poor state of preservation.

Tenon pegs

Acacia pegs secured tenons within their channels. The starboard side contained 121 pegs in total (Figure 5.25). Pegs that were round in section dominated in the construction, but small numbers of rectangular (six) and triangular pegs (two) were also present. The diameter of pegs varied from 20 to 55 mm, but the majority of them had a standard diameter of 35 mm. It will be shown below that pegs were used to secure the ends of the tenons. Thus the majority were placed in double-size channels and the pegs were used in pairs, with each peg securing its tenon. In this case they were spaced across the width of the plank to reduce the risk of splitting it.

In the transversal axis, the distance between the centres of pegs varied from 4.5 to 14 cm, depending on the width of the plank. Many planks showed cracks caused by an insufficient distance between pegs across the width of the plank. It is possible that they appeared after the hull was put under greater strain during its deposition on the sea floor (Figure 5.26). In the longitudinal axis, the distance between paired pegs changed from 10.5 to 19 cm, with an average value of 13.9 cm.

Pegs were driven from the inside of the hull and were visible on its outer surface. While the majority of pegs were levelled with the inner surface of the planking, there were several exceptions that projected 6–10 cm inside the hull (Figure 5.27). These pegs most likely secured elements of internal structure that were not preserved. Their distribution does not allow any determination of the pieces that were secured by them, but in many cases they were probably the bracing timbers.

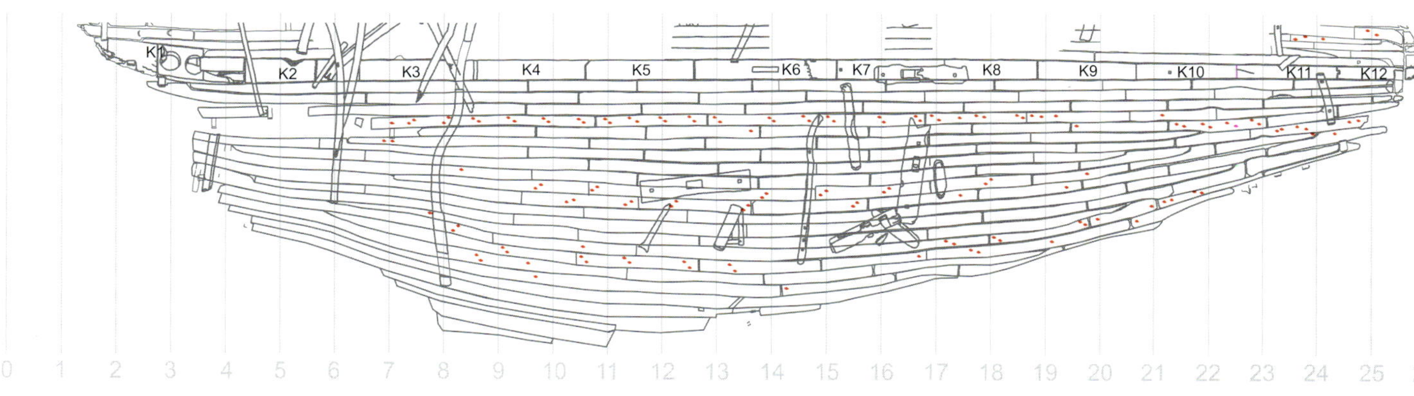

Figure 5.25 Distribution of tenon pegs on the starboard. (Drawing: P. Sandrin/A. Belov; © F. Goddio/IEASM.)

Table 5.9 Width and thickness of tenons of the planking and of wedges preserved inside the channels of the upper preserved strakes (cm). (Distance along the baseline).

Position, m.	**25**	**24**	**23**	**23**	**23**	**22**	**22**	**22**	**21**	**21**	**20**	**20**	**20**	**19**	**19**	**18**	**18**	**18**	**17**	**17**	**17**	**17**	**16**	**16**	**15**	**15**
Strake no.	5	15	15	15	15	16	16	16	18	18	18	18	18	18	18	18	18	18	19	19	19	19	19	19	20	20
Width, cm	16	11	20	14	16	11	12		10	8	13		16	9.5	11.5	13.5		17.5	12.5		12.5		19.5	20	15	8.5
Thickness, cm	3	1.5	3.5	5	5	3	2.5		3	4	4		4	3	3	4		4.5	4.5		4.5		5	5	5	2.5
Wedge, width								?				5					4.5			6		4				
Wedge, thickness								2.5				4					2.5			4.5		3.5				

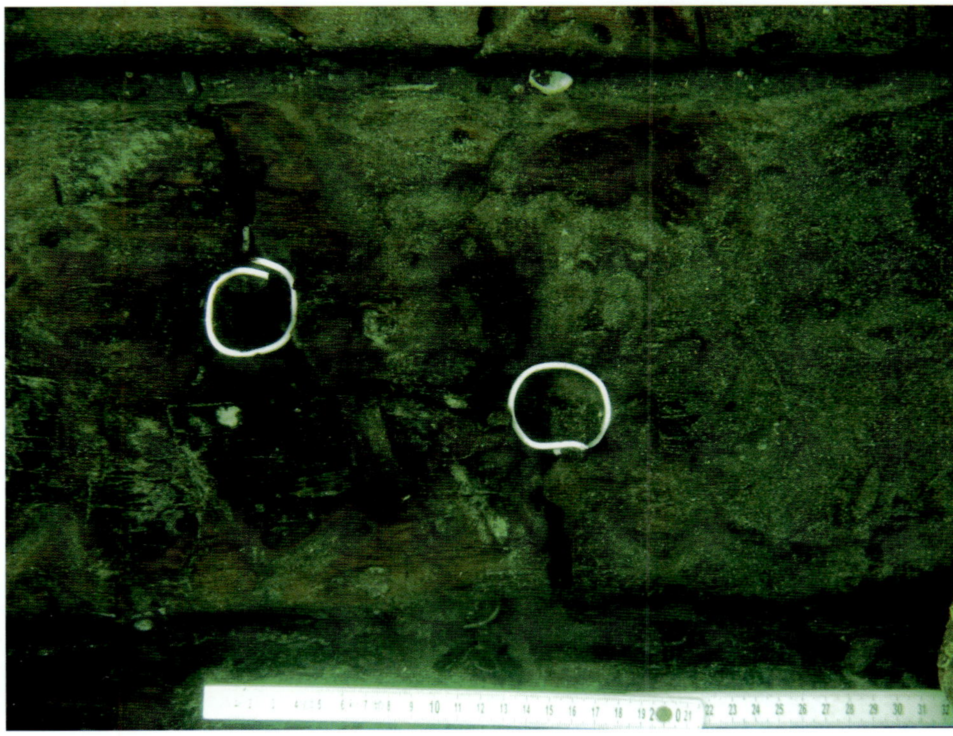

Figure 5.26 Cracks between the pegs of strake S4 (14 m along baseline). (Photo: A. Belov; © F. Goddio/IEASM.)

Figure 5.27 Tenon pegs protruding inside the hull. Left: the strake S10 (10 m along the baseline). Right: the strake P13 (1–2 m along the baseline). (Photo: left: A. Belov; right: C. Gerigk; © F. Goddio/IEASM.)

14	**13**	**13**	**12**	**12**	**11**	**10**	**10**	**9**	**9**	**8**	**8**	**7**	**7**	**7**	**7**	**6**	**6**	**6**	**6**	**6**	**5**	**5**	**5**	**5**	**5**	**4**	**4**	**Average**
19	20	20	20	20	23	23	23	23	23	23	23	18	19	19	19	16	16	16	16	16	3	3	3	9	9	1	2	**cm**
13	12	12	17.5	11	11		10	13	13.5	13.5	13	13	11		12.5	9	10.5	9.5	7.5		12.5		14	8.5	10.5	17	14	12.8
3	4	5	3.5	3	2.5		3.5	4.5	4.5	4	4	3.5	3.5		3.5	2.5	3.0	3.5	3.5		3.5		3	3	3	3.5	3.5	3.633
						4								6						5.5		4						4.875
						4								3						3.5		3.5						3.444

Table 5.10 Number and percentage of pegs in the starboard strakes.

Strake	1	2	3	4	5	6	7	8	9	10	11	12	13	14	15	16	Σ
No. of pegs	0	0	0	44	10	2	6	0	1	14	13	0	0	1	15	15	121
Percent	0.0	0.0	0.0	36.4	8.3	1.7	5.0	0.0	0.8	11.6	10.7	0.0	0.0	0.8	12.4	12.4	100

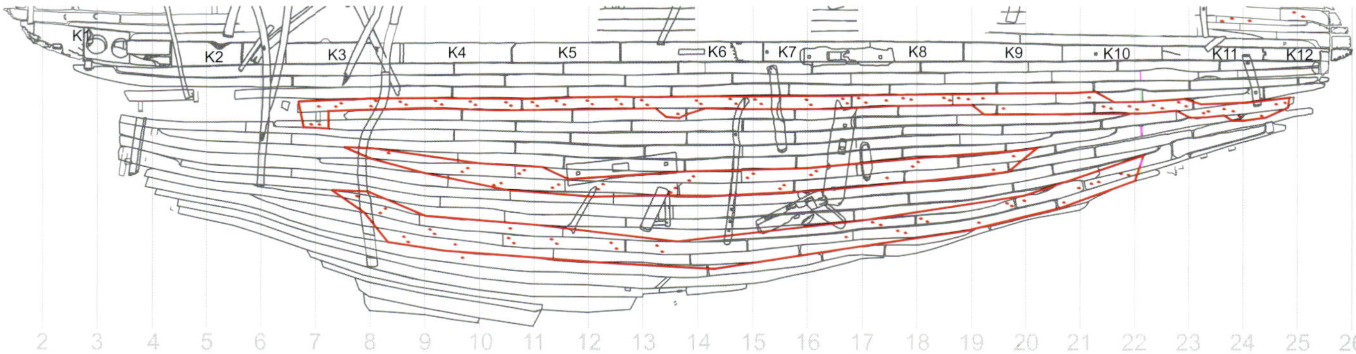

Figure 5.28 Three zones where pegs were concentrated on the starboard of ship 17. (Drawing: A. Belov; © F. Goddio/IEASM.)

Discussion

The manner in which the planking was assembled demonstrates an archaeologically undocumented type of construction, the only parallel of which is in Herodotus' description.[22] The planks were joined transversally using very long pieces of wood that we may call 'tenons'. These long tenons passed through the keel and could assemble as many as five strakes either side of it, i.e. 10 strakes and the keel. The extremities of the tenons were secured to the strakes with pegs, which means that the position of the tenons inside the planking may be inferred from the distribution of the pegs.

Unlike the Greco-Roman type of construction, where pegs were evenly distributed on the surface of the hull, the pegs in ship 17 were concentrated on specific strakes (Table 5.10 and Figure 5.28). Starboard strake S4 contained more than 30 per cent of all the pegs, and strakes S10 and S15 contained 15 per cent each. It is possible to distinguish several zones where pegs were concentrated. The first concentration included starboard strakes S4 and S5, the second was situated in strakes S10 and S11,

Figure 5.29 Tenon that passed through the keel and the garboard of the ship (5 m along the baseline). (Photo: C. Gerigk; © F.Goddio/IEASM.)

Figure 5.30 Double-size channel with the remains of two pegged tenons that supersede one another. (Photo: C. Gerigk; © F. Goddio/IEASM.)

and the third one in strakes S15–S16. The position of these peg-zones corresponds to the length of the tenons. The first tenon passed through the channels of the keel and through those of the first four to five strakes on each side of the vessel. In the region of transition from one tenon to the next, the tenon was superseded by another one that continued in an outboard direction through the strakes from 4–5 to 10–11 (Figure 5.29). A third transversal tenon joined the strakes from 10–11 to 15–16. In transitional areas, double-size channels with average dimensions of 31 × 5.2 cm contained both the outboard extremity of one tenon and the inboard extremity of another. The tenons were thus superimposed inside two neighbouring strakes (Figure 5.30). In this area each tenon was secured with a peg, which is why pegs usually appear in pairs.

A term of 'tenon-rib' has been suggested for tenons used in this type of construction.[23] The awkwardness of this term may be compensated by the presence of the word 'rib' that suggests a parallel with a living body. The planking of the Uluburun ship (*c.* 1300 BC was assembled with long tenons,[24] suggesting tenon-ribs were of primary importance for the transversal structure of ship 17.[25] It would appear, however, that this may not have been the case,[26] as the tenon-ribs are too thin to significantly increase the transversal stiffness of the hull.[27] Consequently, it is now suggested that the tenon-ribs were to fasten the hull rather than to provide additional strength to it.

An interesting question connected with the origin of this constructional technique is whether it is a result of foreign influence or was a purely Egyptian technological solution.[28] Many constructional features of ship 17 strongly indicate that the ship was a freighter of a multipurpose type, built in a local shipyard and out of local wood, suggesting that this was probably a local form of ship construction.[29]

Lashings were frequently used in the Mediterranean boatbuilding tradition during the period from the eighth to fifth centuries BC and they were often applied together

22 See this volume, Chapter 10.
23 Belov 2014a: 184.
24 Pulak 1988; 2008: 303. The tenons of this ship never exceeded their own strake.
25 Belov 2014b: 318; 2015a: 199.
26 The revision of this writer's view is due to fruitful discussions with J.-P. Olaberria, to whom I express my sincere gratitude. The same ideas can be found in Barker 2014: 479, 481.
27 Second moment of area *I*, otherwise called 'moment of inertia of a section', of a tenon is calculated by the following formula: $I = a \times b^3/12$; where *a* is the width of a tenon and *b* is its thickness. Thus the inconsiderable thickness of tenons of ship 17 does not significantly increase the resulting value of the moment of inertia, which remains very low.
28 Discussions of the important innovations that appear in Egyptian shipbuilding during the eighteenth dynasty go beyond our subject. Different opinions can be found in Köster 1923; Glanville 1930; 1932; Février 1935; Faulkner 1940; Säve-Söderbergh 1946; Helck 1962; Landström 1970; Basch 1978; 1987; Ward 2000.
29 See this volume, Chapter 10.

Table 5.11 Dimensions of mortises in the planking of ancient Egyptian ships (cm).

Ship/timbers	Date	Mortises (width × thickness)	Ligatures
Abydos	2650 BC	7–8 × 1.7–2.0	yes
Lisht	1950 BC	9–10 × 1.5	yes
Dahshur	*c.* 1878–1841 BC	7.5 × 1.8–2.0	no
Mersa Gawasis	*c.* 1956–1650 BC	8–9.5 × 1.5–1.8	yes
Ayn Sukhna	*c.* 2500–2400, 2000–1800 BC	single: 15 × 6–8.5 × 1–2 double: 15 × 5.5–8.3 × 1.1–2	yes
Mataria	*c.* 450 BC	14 x 3.1	no

Table 5.12 Dimensions of tenons in the planking of ancient Egyptian ships (cm).

Ship	Date	Tenons (long, wide, thick)	Ligatures
Khufu I	*c.* 2500 BC	10 × 7.0 × 1.5	yes
Lisht	*c.* 1950 BC	[10.5–11.5] × 6.5–4.5 × 1.5	yes
Dashur	*c.*1878–1841 BC	22–25 × 7.5 × 1.8	no
Mersa Gawasis	*c.* 1956–1650 BC	11 × 4.0 × 1.0	yes
Ayn Sokhna	*c.* 1991–1649 BC	15 × 5.5–8.0 × 0.9–1.9	yes
Mataria	*c.* 450 BC	> 17 × 10 × 2.0	no
Kyrenia	4th cent. BC	15 × 4.3 × 0.5	no

with a mortise-and-tenon joinery.[30] Egyptian boat builders also used lashings in the construction of their vessels from the First Dynasty, as seen in the Abydos ships (2650 BC).[31] It appears, however, that this technology developed independently in Egypt, with the direction of lashing being different from that of Mediterranean ships.[32] The use of lashing can be traced archaeologically until the end of the Middle Kingdom,[33] but there is an important lacuna in our knowledge for later periods. The constructors of the Mataria boat (c. 450 BC) relied on mortise-and-tenon joinery only and no traces of lashings were found during the excavations of ship 17.

Mortise-and-tenon joints were used in boatbuilding from at least the Old Kingdom[34] and Egyptian shipwrights used them together with lashings in every boat in the archaeological record, with the possible exception of the aforementioned boat timbers from Abydos. According to Ward, the large tenons of the Mataria ship continue an ancient tradition.[35] Vinson, relying on archaeological material, suggested that the tenons of Egyptian ships were very large by Greek standards.[36] Tables 5.11 and 5.12 show the dimensions of mortises and tenons from the timbers of Egyptian ships and also the size of tenons of the Kyrenia ship (fourth century BC), which is used as a roughly contemporary example of the ancient Greek boatbuilding tradition.

With the exception of the tenons from the planks from Mersa/wadi Gawasis,[37] the tenons of ancient Egyptian ships are all larger than those of the Kyrenia ship. This difference cannot be explained by the Egyptian vessels being bigger; the Kyrenia ship was larger than the Dahshur and Mataria ships.[38] Furthermore, the tenons from the planks from Lisht and from the Mataria ship were not completely preserved and their initial length would have been even more considerable.[39] The latter is particularly pertinent as it was nearly contemporary with ship 17 from Thonis-Heracleion. According to Ward, some planks gave 'an idea of the enormous tenons used to join plank edges'.[40]

It is true that the tenons used in all of these vessels were used simply to join adjacent planks, rather than multiple runs of them, as was the case in ship 17. But this does not necessarily mean that similar vessels with long tenon-rib construction were not built earlier: archaeological data on Egyptian shipbuilding is too fragmentary to allow such a definite conclusion. In this light, some

30 McGrail 2008: 618.
31 Ward 2003.
32 Ships belonging to the Mediterranean type of construction were lashed along the seams between the strakes, while the Egyptian ones were lashed in the transversal axis. The best examples are provided by the Khufu I ship (*c.* 2500 BC) and the Abydos boats (*c.* 2650 BC).
33 Lashings were found in timbers from Lisht (*c.* 1950 BC), from Mersa/wadi Gawasis (while the site is difficult to date with precision, its most intensive occupation corresponds to the Middle Kingdom [*c.* 2055–1650 BC], see Ward and Zazzaro 2010) and in the construction of ships from Ayn Sokhna (*c.* 2500–2400, 2000–1800 BC, see Pomey 2012a; 2012b).
34 This type of joinery was known to Egyptians as the 'Teeth of Osiris': *Book of the Dead*, Ch. 99, Book 18, see Montet 1925. Incontestable iconographic and archaeological proofs are provided by the boatbuilding scenes in the tomb of Ty (Fifth Dynasty, *c.* 2494–2345 BC), which are the earliest representations of mortise-and-tenon joinery in the Mediterranean, and by the Khufu I ship (Fourth Dynasty, *c.* 2550 BC), where the role of mortise-and-tenon joinery seems to be underestimated by scholars: see Ward 2000: 49.
35 Ibid.: 133.
36 Vinson 1998: 256.
37 Planks from Mersa/wadi Gawasis were quite specific as the mortises formed two rows across the planks thickness and there were also dovetails present, see Ward and Zazzaro 2010. Double rows of mortises were equally attested in planks from Lisht, see Haldane 1993: 237. Another example of this technique is provided by the material from Ayn Sokhna. See Pomey 2012a; 2012b.
38 Vinson 1998: 251–60.
39 See Haldane 1993: 237; Ward 2000: 133.
40 Ibid.

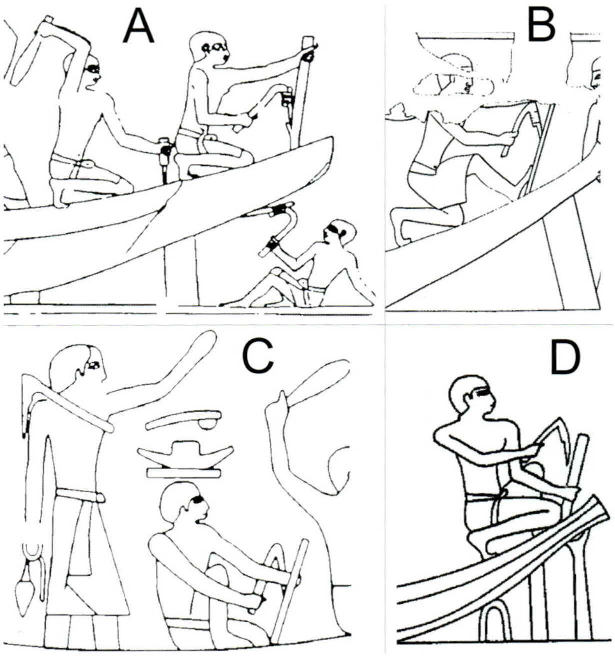

Figure 5.31 Old Kingdom reliefs showing workmen cutting long pieces of wood: (A) tomb of Ty from Sakkara, fifth dynasty; (B) tomb of Ni'ankh-khnum and Khnemhotep from Sakkara, fifth dynasty; (C) tomb of Khuenre from Giza, fourth dynasty; (D) tomb of Khunes from Zâwyet el-Maiyitîn, fifth dynasty. (Drawing: A. Belov after M. Rogers 1996: figs 13, 17, 21, 38).

details of the boatbuilding scenes in the tombs of the Old Kingdom deserve closer attention. Many of the reliefs show workmen cutting long and narrow pieces of wood (Figure 5.31). These have been interpreted as treenails, tenons, or both.[41] It is interesting to note that other workmen depicted on the same reliefs are occupied with cutting the mortises in the planking.[42] It is important to keep in mind that once there is a physical relationship between two elementary objects (*unico*) on an Egyptian representation, their relative dimensions are always respected.[43] Since there is a relation of scale between the workman and the part he is trimming, it is reasonable to make an estimate of the size of the latter from the size of the former. The object is more than half the height of the workman[44] on the majority of the reliefs and thus can be estimated as at least at 80–90 cm long. If these objects were tenons, they would be for craft smaller than ship 17.

The wedges that secured the tenons of ship 17 within their channels seem to find parallel in 'wooden slips' known from Lisht planks.[45] Apparently the pegging of

Figure 5.32a–b Butt end of one of the planks of ship 17 that forms a part of a half-lap joint (strake P14, 6 m along the baseline). (Photo: A. Belov; © F. Goddio/IEASM.)

tenons was not typical of Egyptian shipbuilding before the Late Period.[46] Even in those known cases[47] when Egyptians did peg the tenons it was done in a slightly different way to the Syro-Canaanite and Greco-Roman boatbuilding traditions, although the principle remained the same. According to these traditions, every joint was pegged on either side of a plank seam and pegs were visible from the outside of the hull, whereas the Egyptians applied pegs only to some selected joints.[48] Pegged tenons used for repairs of the planking (before *c.* 500 BC) never went through the plank's thickness.[49] The ends of tenons of the planking of ship 17, however, were regularly secured with pegs that went through the plank's thickness. This implies an important change in boatbuilding technology towards the Late Period.[50]

41 Rogers 1996: 68–74.

42 It is peculiar that mortise-cutting scenes are found only on the reliefs of the tomb of Ty: ibid.: fig. 6A.

43 Vergnieux 1990.

44 The average height of an ancient Egyptian male in this period is estimated to be 1.65 m (see Stock *et al.* 2011: 355, table 14.2).

45 Haldane 1993: 172.

46 Tenons of the Khufu I ship, those from Mersa/wadi Gawasis planks and from the ships of Ayn Sokhna were unpegged (free tenons). In two latter cases planks belonged to seagoing ships and free tenons allowed easy assembly/disassembly of hulls for their transportation or storage. See Ward and Zazzaro 2007; Ward 2009; Pomey 2012a; 2012b.

47 The Mataria ship (*c.* 450 BC). Regular use of tenon pegs seems to be unfamiliar to ancient Egyptian boatbuilding before *c.* 500 BC. See Ward 2004: 18.

48 Ward 2000: 133–4.

49 Ibid.: 98–100, 115, 119. However, Wachsmann cites a representation from the tomb of Qaha at Deir el-Medina (end of the nineteenth dynasty, 1295–1186 BC) that seems to show the pegs of the planking. Wachsmann 1989: 199.

50 The most ancient evidence for regular use of tenon pegs is provided by the Uluburun ship (*c.* 1300 BC). It is thus theoretically possible that Egyptians borrowed this method from the Levant.

Figure 5.33 Half-lap joint in the upper preserved strake of ship 10 from Thonis-Heracleion (^{14}C calibrated date: 475 ± 45 BC. See Goddio 2007: 106). (Photo: P. Rousseau; © F. Goddio/IEASM.)

Figure 5.34 Position of the tenon-ribs inside the outer preserved strakes of ship 17 (bow starboard area). (Photo: C. Gerigk; © F. Goddio/IEASM.)

The longitudinal joints of the planking

The majority of planks composing each strake of ship 17 were assembled with a half-lap joint (Figures 5.32a-b–5.33).[51] The overlap ranged between 20 and 45 mm.

It is interesting to note that the rather dense distribution of tenons, along with their corresponding channels, allowed their regular placement near to the ends of planks. This would have provided extra support to the joint between the planks of the same strake (Figure 5.34).[52] Several planks were assembled with more complex scarfs judging by their broken line in sheer-view (Figure 5.35).

The more complex Jupiter scarf was applied in the construction of Khufu I ship, while joggles were characteristic of the planks from Lisht belonging to a Nile freighter.[53] Vertical lines between the ends of the planks on ancient Egyptian representations can be interpreted as a butt joint or a half-lap joint. It was probably this last joint that the shipwrights used in the planking of the Mataria ship (*c.* 450 BC).[54] An ethnographic parallel for

51 This is another parallel with the construction of the Mataria boat (*c.* 450 BC). See Ward 2000: 131.

52 Another technique was applied in the Lisht planks where the mortises of the planking occupied the middle of the joint. These were classified as 'open mortises'. See Haldane 1992: 107, pl.132.

53 Haldane 1992: pl.125, 132; Ward 2004: 13.

54 Ward 2000: 131.

Figure 5.35 Scarfed ends of the planks of strake S13 (starboard, 21 m along the baseline) and the outboard extremity of support F8S (reused plank). (Photo: A. Belov; © F. Goddio/IEASM.)

the use of this joint is seen in the Nile boat called a *naggr* (or *nuggar*) that preserved many features of its ancient predecessors.[55]

Wales and the position of a chine

No pronounced difference in the thickness of strakes was seen, although the data on the planking is not complete. The thickness of strakes measured in the transversal cut in the bow area (strakes S4-S16) changed from 12.7 to 17.5 cm, with strake S16 the thickest (see Table 5.6). The considerable and uniform thickness of the strakes of ship 17 suggests that wales were absent in this type of construction.[56] Thus the position of a chine is difficult to determine. Some supplementary considerations indicate the position of a chine near starboard strake S13.[57]

55 See discussion in the volume, Chapter Four and Clarke 1920: 46; Hornell 1943a: 29.

56 The wales of the Kyrenia ship (fourth century BC), for example, were more than twice as thick as the bottom strakes (8 cm against 3.7 cm). See Steffy 1985: 87.

57 See this volume, Chapter 11.

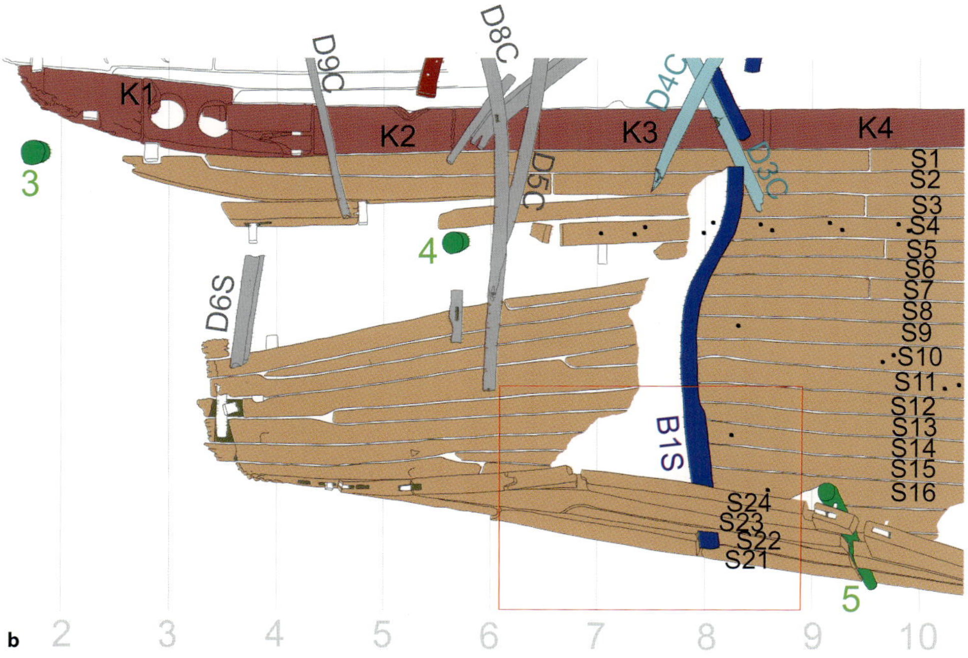

Figure 5.36 Photograph (a) and plan (b) showing strakes S23 and S24 inclined inside the hull. Red rectangle indicates the area covered by the photo. Note the bow of ship 63 at left bottom corner of the photo and the end of through-beam B1. (Photo: C. Gerigk; drawing: P. Sandrin/A. Belov © F. Goddio/IEASM.)

Bulwark

The strakes of ship 17 that were partially preserved above the level of the through-beams were classified as a part of a bulwark.[58] Usually the position of beams cannot be used as a reliable criterion for a bulwark as quite often it was situated above the beams.[59] In Egyptian naval architecture, however, the deck, often in the form of removable hatches, was from the Old Kingdom installed on the through-beams and there are no known exceptions to this rule. Where several levels of through-beams are represented,[60] the deck was probably installed on the uppermost level. Most often the bulwark was composed of only one strake, although the seagoing ships of the expedition of Queen Hatshepsut to Punt were probably equipped with a two-strake bulwark.[61] That was possibly also the case with the warships of Ramesses III (1184–1153 BC), as the bulwark was designed to protect the rowers from enemy archers.

The remains of through-beam B1S on the starboard and of three more through-beams (B1P, B2P, B3P) on the port of ship 17 indicate that strakes S23, S24, P21, and P22 were a bulwark. Like other strakes, they were composed of short planks assembled by a half-lap joint.[62] The only

58 *ḥrit*, *ḥryt*. See Jones 1988: 177, III.110; Vinson 1997. The use of this term remains conventional in the case of ship 17, which in all probability was not decked.

59 For example that was the case of the Kyrenia ship (fourth century BC). See Steffy 1985: 95.

60 Obelisk barges represented in the mortuary temple of Queen Hatshepsut (eighteenth dynasty, 1473–1458 BC) and large river-going ships on the so-called 'blocks of Piankhy' (twenty-sixth dynasty, Psamtik I, 664–610 BC).

61 See reconstructions by Landström 1970: 124–7.

62 The bulwark of the Dahshur boats consisted of 3–4 planks assembled by a butt joint. See Ward 2000: 90.

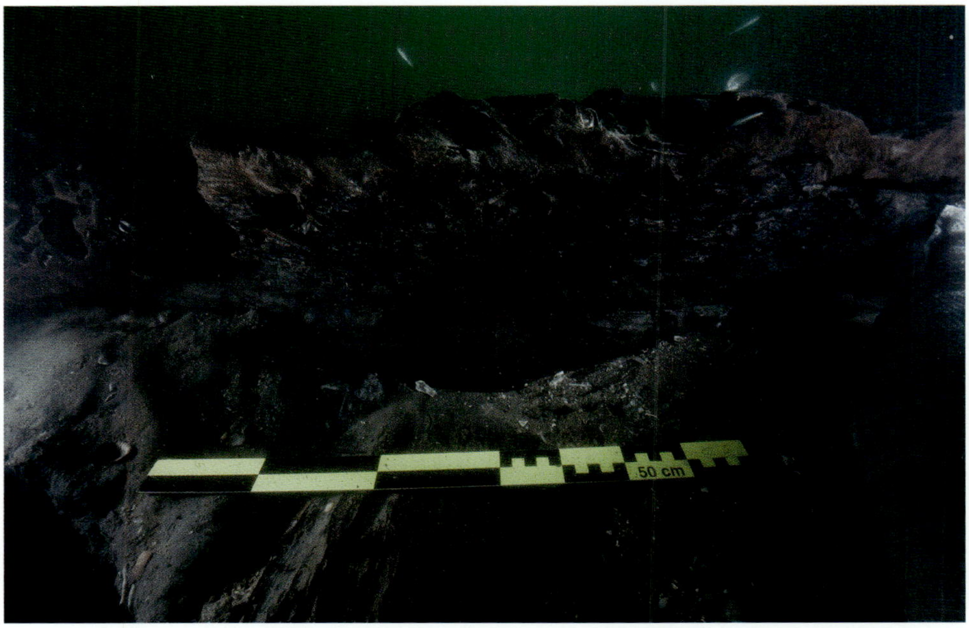

Figure 5.37 Strakes of bulwark S23 and S24 viewed from the interior of the hull. Note the through-beam B1 in the foreground. (Photo: C. Gerigk; © F. Goddio/IEASM.)

difference resides in their flattened and smoothed inner surface as well as their greater width (between 27.3 and 25 cm against 18 cm on average for the other strakes). Their outer surface was poorly worked and was characterised by convex section with a maximum thickness of 16 cm in the middle of the planks width that diminished to 10–11 cm at the edges. The best-preserved segment of the bulwark was 6.4 m long and found at the starboard stern area where it was probably supported by well-preserved beam B1. The bulwark was, however, broken and inclined inside the hull following the impact of the stem of ship 63 (Figures 5.36a-b–5.37). The preserved height of the bulwark reached 41 cm above the upper surface of through-beam B1. No trace of a gunwale was found. The port bulwark, consisting of strakes P21 and P22, was preserved for 19.1 m in the middle part of the hull. Its preserved height was 36 cm.

Haldane suggested that the bulwark played an important role in the longitudinal structure of ancient Egyptian ships as an effective solution for countering the hogging of a sickle-shaped hull.[63] This hypothesis is supported by the Old Kingdom representations on which workmen install bulwark with heavy pounders during the last stages of construction.[64] Vinson comes to the same conclusion after analysing *P.Anastasi* IV, which dates to the Ramesside Period.[65] It seems that the principal reason for the destruction of the boat mentioned in the papyrus was that 'They didn't give it some long bulwarks of acacia, or some long bulwarks of *imw*-wood, so that they could grip the vessel'.

Table 5.13 Width of visible layer of luting between starboard strakes of the ship (cm, position along the baseline).

Strake	Position	Width, cm	
S1/S2	23	1.0	3.5
S2/S3	23	3.0	
S4/S5	23	2.0	
S6/S7	22	2.0	
S7/S8	22	2.0	
S13/S14	21	⇔ 6.0	⇳ 2.5
S16/S17	19	⇔ 5.0	
S17/S18	19	⇔ 3.5	⇳ 2.0
S8/S9	17	2.5	
S19/S20	16	⇔ 6.0	⇳ 3.0
S11/S12	14	⇔ 5.0	
S18/S19	13	2.5	
S4/S5	12	2.0	3.0
S4/S5	11	⇔ 3.0	⇳ 3.0
S7/S8	8		⇳ 6.0

Legend:
⇔ 6.0 width of luting layer between the butt ends of the planks (same strake)
⇳ 3.0 width of luting layer between the planks edges (neighbouring strakes)

The upper strake of ship 17 was only partially preserved and no traces of any potential rowing arrangement were found.[66] It is quite possible that ship 17 was not oared. Herodotus does not mention oars in his description of the *baris*. In case of unfavourable wind the *barides* were towed upstream.[67] It should also be noted that no oars are seen in the depiction of a *baris* in the Palestrina mosaic.[68]

63 Haldane 1993: 234–5.
64 Representations from the tomb of Ty from Saqqara (fifth dynasty, 2494–2345 BC). The length of bulwark was estimated at 10 m (Boreux 1925: 244) and it is quite probable that it consisted of several planks. See Rogers 1996: 59.
65 Vinson 1997: 158.
66 Usually ancient Egyptians used grommets (oar loops) that were reeved through mortises in the planksheer or bulwark. See Boreux 1925: 338; Doyle 1998: 166–75.
67 Herodotus *Historia* 2.96 lines 10–3.
68 Pomey 2015b: 165–6 and note 43.

Figure 5.38 Luting of the joint between the planks of starboard strake S13 (21 m along the baseline). (Photo: A. Belov; © F. Goddio/IEASM.)

Luting of the planking

Traces of vegetal material[69] in the form of strands driven between the seams of the planks were distinctly seen in both the longitudinal and transversal joints *inside* the planking of the ship (Figures 5.38–5.39). This material was visible only in the pronounced gaps between the plank edges that varied in width from 1 to 6 cm, with an average value of 3.3 cm for the starboard (Table 5.13). Normally the gaps between the butt edges of the planks belonging to the same strake were more pronounced than those between their side edges.

The choice of the term 'caulking' would not be correct as it suggests that the material is driven by force from the outer side of the hull.[70] This was not the case with ship 17, where the gaps between the seams were filled from the inside; thus the term 'luting' is preferred.[71] The same luting technique is used in the planking of the *nuggar*, the only difference being that old rags are used instead of vegetal material.[72]

Luting was applied only to some selected joints of the planking of ship 17 probably to make watertight joints that were either defective or complex in configuration.[73] No particular pattern in the distribution of the luting

69 The analysis of this material has not yet been undertaken.
70 An exhaustive analysis of the meaning of the term is proposed in Basch 1986; 2008; 2015.
71 Luting means the application of waterproofing material between the joints of the planking in progression with the building process: see Steffy 1998; Pomey and Rieth 2005: 212.
72 Clarke 1920: 50; Hornell 1943a: 29–30.
73 The same phenomenon was characteristic of the Dramont 'E' shipwreck from Saint-Raphaël (fifth century AD). See Santamaria 1995: 149.

Figure 5.39 Close-up view of the same joint. (Photo: A. Belov; © F. Goddio/IEASM.)

was apparent, although it seems that more luting is to be found at the extremities of the vessel, as these contained planks that were complicated in shape.

Miscellaneous parts of construction found on inner planking: the vestige of a deck?

Normally, ancient Egyptian ships were decked.[74] This conclusion is supported by numerous representations of river-faring boats transporting goods on their decks that appear from the Old Kingdom.[75] Decking consisting of removable covering boards was a common type,[76] which could be removed to allow rowers sitting on through-beams to lower their legs into spaces in the hull.[77]

It is generally suggested that cargo was transported on the deck of ancient Egyptian boats in order to distribute its weight on the larger surface of the hull by means of through-beams, for example, on Hatshepsut's boats at Punt.[78] The other reason for such a solution might have been that the hull was not watertight, although it is possible that Egyptian artist may have simply wanted to show the cargo that in reality was stowed in the hold.[79]

74 Boreux 1925: 307–10, 333; Bell 1933: 101–11; Landström 1970: 60; Rougé 1975: 43, 77–8; Jones 1990: 54.

75 See Landström 1970: 60, 122–3, 134–5, 138, figs 175–80, 372, 389–96, 405.

76 Removable deck first appears in the models of the twelfth dynasty; however, it is quite probable that it existed before that date. See Boreux 1925: 307–10, 333; Landström 1970: 78, 100.

77 Boreux 1925: 333. Boreux proposes a parallel with modern Egyptian *dahabiyehs* and *beden-seyads* from the coast of the Arabian peninsula.

78 See Bell 1933: 101–11; Jones 1990: 54; Landström 1970: 60; Rougé 1975: 43, 77–8; Jones 1995: 43.

79 Seagoing ships of Queen Hatshepsut returning from Punt may serve as an example of this practice.

A small hatch or a door was discovered inside the hull of ship 17 (Figures 5.40–5.41).[80] It was built from six planks and was 66 cm long, 46 cm wide and had a changing thickness between 1.5 and 2 cm (Figure 5.42). Mortise-and-tenon joints between the planks were secured by two small pegs on one of the short sides. Traces of a mortise, probably destined to contain a hinge, were still visible in the middle of one of the long planks.

Beams from the Khufu I ship, the Dahshur boats and the beam T32 of Mersa/wadi Gawasis had ledges on their upper surface to receive deck-planking.[81] This is not the case with the preserved through-beams of ship 17, which were inclined and had irregular forms that would not have permitted the installation of a deck.[82] The hatch indicates that at least the stern of ship 17 was possibly decked and fitted with a small cabin with a door and a platform to facilitate steering: a helmsman's platform.[83] The *baris*-boats depicted on the Palestrina mosaic also have a deckhouse located aft of the mast.[84]

The absence of a deck did not hinder the capacities of ship 17 as a freighter. Her proto-keel and massive planking would easily have allowed the transport of goods in the hold. It is notable that Herodotus, in his description of the *baris*—a type to which ship 17 most probably belonged—does not mention a deck. This omission, in what is otherwise a highly-detailed text, testifies to the absence of deck in construction of the *baris* that he observed.

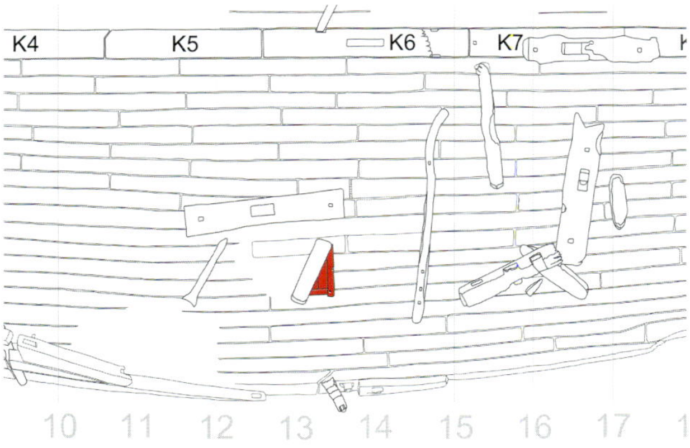

Figure 5.40 Position of a hatch or a door on the plan of ship 17. (Drawing: P. Sandrin; © F. Goddio/IEASM.)

Figure 5.41 Photo of a hatch or door. (Photo: C. Gerigk; © F. Goddio/IEASM.)

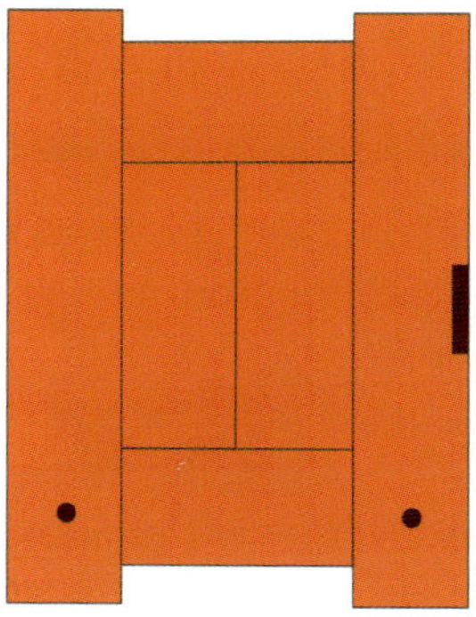

Figure 5.42 Schematic plan of the construction of the hatch or door. (Drawing: A. Belov.)

80 Cf. Casson 1995: 134 discussing Greek 'hollow' ships that were decked at the extremities of the hull only.

81 Ward and Zazzaro 2010: 32–3.

82 Indirect evidence comes from the other ships of the Late Period from Thonis-Heracleion. Large blocks of limestone were discovered inside hulls of some ships obviously belonging to the same constructional type as ship 17 (e.g. ship 5, dating to the fourth–second century BC). Here it is not the place to argue whether these ships were transporting the blocks, whether they were ballast or were scuttled with their help; the important thing is that there are no traces of deck under the blocks that lie directly on the planking.

83 According to the 3D reconstruction of ship 17, its stern containing the rudder shafts is characterised by a pitch of around 30°. Thus the helmsman needed a more horizontal area such as a steering platform or the roof of a deckhouse. Either possibility cannot be excluded: see Belov 2014c: 8.

84 Pomey 2015b: 165.

Figure 6.5 Bracing timbers of different sections. A: F3P, octagonal; B: F10S, hexagonal; C: F1P, oval; D: F6S, rectangular). (Photo: A: A. Belov; B–D: C. Gerigk; © F. Goddio/IEASM.)

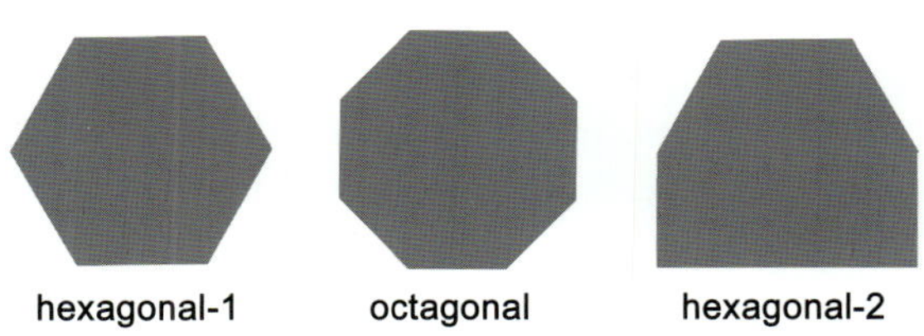

Figure 6.6 Different kinds of polygonal form in the sections of bracing timbers of ship 17.

Table 6.1 Dimensions and shape of bracing timbers.

Bracing timber	Length, cm	Form in section	Width (avg), cm	Thickness (avg), cm	Number of fastenings
F10S	99	hexagonal 2	16	8	2
F9S	71	rectangular	20	7	0
F7S	167	rectangular	16	10	0
F6S	274	rectangular	12	7	4
F10P	187	octagonal	11	7	6
F9P	145.5	rectangular	18	7	0
F6P	262	hexagonal 1	12	7	3
F3P	275	octagonal	13	10	9
F2P	221	octagonal	12	10	5
F1P	328	oval	18	7	9 (4 circular)

Dimensions

Information on the preserved length, average width, average thickness, dominant form in section, and number of pegs for the preserved bracing timbers of ship 17 is presented in Table 6.1.

Fastenings

Bracing timbers were attached to the planking by mortise-and-tenon joints. The majority of the tenons (38) were rectangular in section but there were also four that were circular.[1] All of the preserved tenons projected 1–9 cm above the upper surface of the bracing timbers (4 cm on average) (Figure 6.7). This could suggest that other timbers were attached on top of them, which recalls the 'bulkheads' from the Lisht ship.[2]

Tenons were distributed unevenly along bracing timbers and the distance between them varied from 14 cm to 1.39 m (48.9 cm on average) (Figure 6.8). Tenons were uniform in shape, on average 4.6 cm wide and 3.9 cm thick.

1 All of these belonged to bracing timber F1P.

2 Haldane 1996a.

Figure 6.7 Three tenons inside the mortises of bracing timber F6S. (Photo: A. Belov; © F. Goddio/IEASM.)

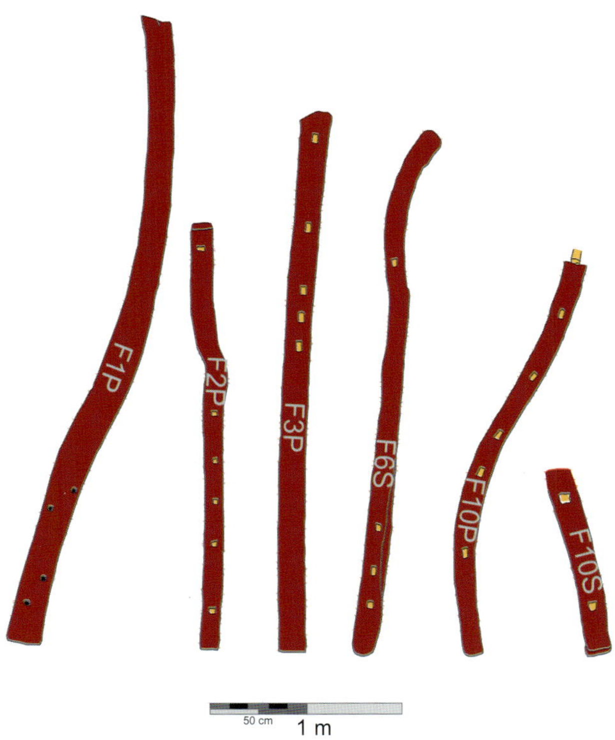

Figure 6.8 Distribution of fastenings in selected bracing timbers. (Drawing: P. Sandrin/A. Belov; © F. Goddio/IEASM.)

Supports

Five short massive elements of the internal structure of the ship were classified as 'supports' (Figure 6.9). Most of them were oriented at 90° to the keel and probably served to distribute the weight of the beams across the surface of the planking by means of stanchions.

Position

Supports occupied the central area of the hull from 12 to 17 m along the baseline (Figure 6.9). Supports F8P, F8S, and F5P were oriented at 90° to the keel. The distance between the side of the keel and their nearest extremities varied from 60 to 70 cm.[3] The fourth support (F5S) was found in a position that was almost parallel to the keel, however, it originally had the same position as the others. The fifth support (D1C) was oriented along the keel and placed above it.

Shape

Each support was in the form of a cuboid (Figures 6.10–6.15). Several peculiarities of shape may be noted. The keel-side extremity of support F8S ended with two triangular teeth, while F8P had bulges at both extremities. It will be shown below that the supports were fabricated out of reused planks and thus the shape corresponds to a scarf of the longitudinal assemblage Each piece also had from three to four channels cut through their sides that initially served for the transversal assembly of the planks.[4] A large rectangular mortise was cut in the middle of the upper surface of each support (Figure 6.16).

3 Exact distances were as follows: F8P – 59.2 cm; F5P – 60.8 cm; F8S – 67.8 cm.

4 Supports F5P, F8S, and D1C had three channels while F5S and F8P had four.

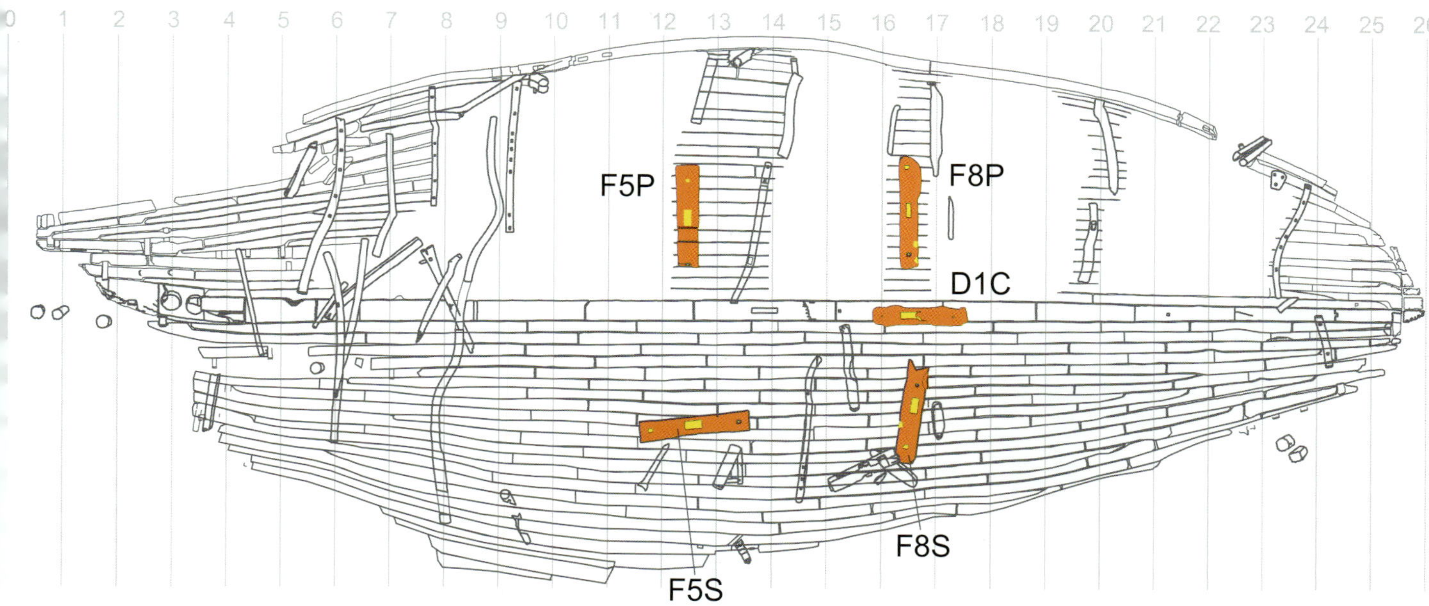

Figure 6.9 Position of supports F5S, F5P, F8S, F8P and D1C on the general plan of the ship. (Drawing: P. Sandrin/A. Belov; © F. Goddio/IEASM.)

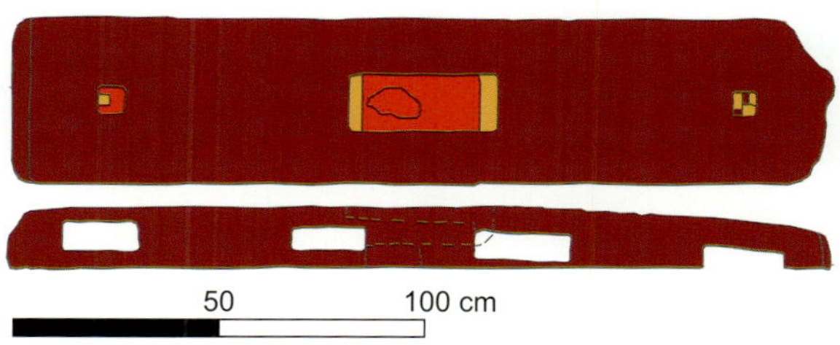

Figure 6.10 Plan and profile of support F5S. (Drawing: P. Sandrin/A. Belov; © F. Goddio/IEASM.)

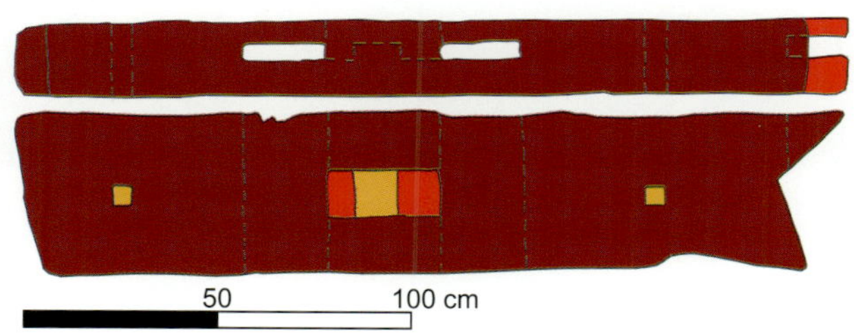

Figure 6.11 Plan and profile of support F8S. (Drawing: P. Sandrin/A. Belov; © F. Goddio/IEASM.)

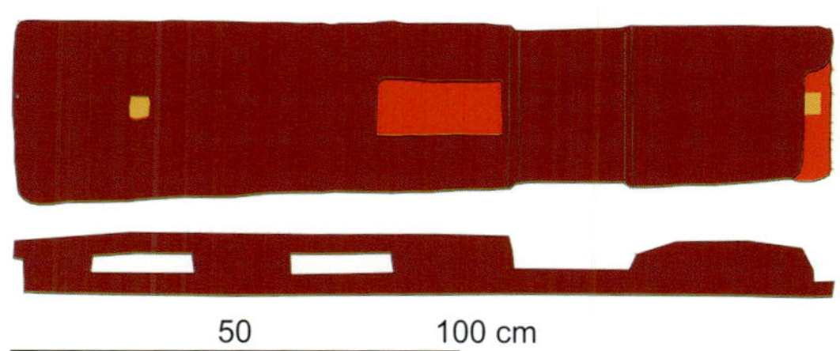

Figure 6.12 Plan and profile of support F5P. (Drawing: P. Sandrin/A. Belov; © F. Goddio/IEASM.)

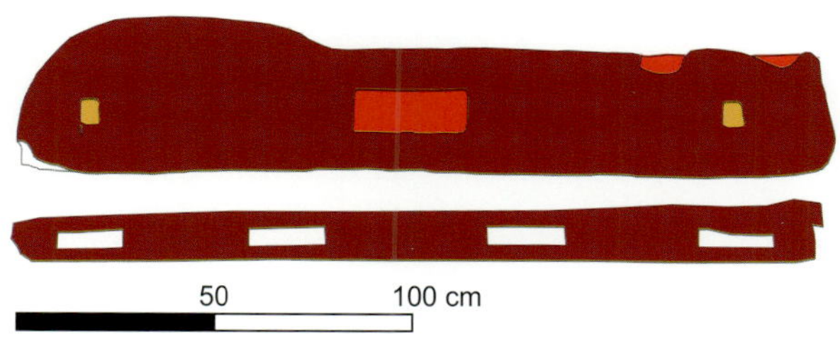

Figure 6.13 Plan and profile of support F8P. (Drawing: P. Sandrin/A. Belov; © F. Goddio/IEASM.)

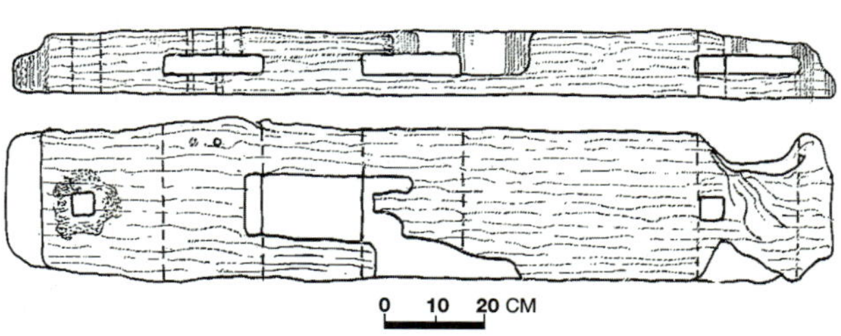

Figure 6.14 Plan and profile of support D1C. (Drawing: P. Sandrin/A. Belov; © F. Goddio/IEASM.)

Figure 6.15 Photomosaic of support F8P. (Photo: C. Gerigk; © F. Goddio/IEASM.)

Figure 6.16 Mortise cut in the upper surface of support F5S. (Photo: C. Gerigk; © F. Goddio/IEASM.)

Dimensions

The main dimensions of the supports are shown in Table 6.2. They have a length to width ratio of about 1:5 and length to thickness ratio of 1:10. Mortises of supports F5S, F8S, and F8P contained the remains of vertical pieces of wood (Figure 6.17). The dimensions of upper mortises and lateral channels are shown in Tables 6.3 and 6.4.

Table 6.2 Main dimensions of supports.

Supports	Length, cm	Width, cm keel-side extremity	Width, cm outboard extremity	Thickness, cm keel-side extremity	Thickness, cm outboard extremity
F8S	211.4	40	41.9	19	18.1
F5S	199.5	39.2	38.6	16	11.5
F8P	206	32	39	14	10
F5P	183	34.6	40.6	6.7 (not preserved completely)	13.1
D1C	164	34.4 (stern extremity)	29	11 (bow extremity)	13

Figure 6.17 Remains of a vertical piece of wood inside mortise of support F8S. (Photo: C. Gerigk; © F. Goddio/IEASM.)

Table 6.3 Dimensions of mortises cut in the upper surface of supports and information on remains of vertical pieces of wood inside them.

Support	Length, cm	Width, cm	Depth, cm	Remains of details inside mortises
F8S	27	10.5	7	one (13×10×7)
F5S	30	14	5	two (14×4×5)
F8P	29	10.5	5	no
F5P	28	13	5	traces
D1C	26	12.5	5–7	no

Fastenings

The end of a support was attached to the planking by vertical tenons installed in square mortises with a side of about 5 cm. With the exception of one of the extremities of F5S, which was secured by three small tenons, only one tenon was employed at each end. The associated tenons were square in section with a side of 3.9 cm on average. A supplementary tenon, 7 × 4.5 cm in section, was found in the side of F8S (Figure 6.18). An opening of the same dimensions in the side of F8P suggests that it too once contained a tenon.

Bracing timbers and supports: general conclusions

Bracing timbers and supports were undoubtedly installed after the assembly of the planking. This is corroborated by the thick planking of ship 17 and the relatively thin and asymmetric bracing timbers that could in no way serve as frames to support it. As will be shown later, the supports were installed before or at the same time as the beams and definitely preceded the installation of the bulwarks.[5]

The size and shape of the supports, as well as the dimensions and distribution pattern of the channels within them, demonstrate that they were reused planks. The ends of the supports were shaped to form part of a half-lap or scarf joint for their longitudinal assembly within a strake (F5P, F8P and F8S). The length of the supports varied from 1.64 to 2.11 m and corresponds well with the average value of 1.92 m established for the starboard planks of ship 17. The thickness of the supports (10–19 cm) was also quite similar to that of the planks (10–18 cm). The supports, however, were generally larger (29–41.9 cm) in comparison to the starboard planking (18.8 cm on average and never exceeding 29 cm). It is possible that supports were chosen from planks that were formerly part of a bulwark, which were normally larger than ordinary strakes. The form and dimensions of the channels cut in the sides of the supports closely follow those of these wider planks.[6]

Table 6.4 Dimensions of lateral channels cut inside the supports.

Mortise No.	Dimensions	F5P	F5S	F8P	F8S	D1C
1	Length, cm	25.5	19	19	20.5	20
	Width, cm	[?]	7.5	5	5	5
	Distance	52	55.5	54	51	48.3
2	Length, cm	23	18	19	21	32.5
	Width, cm	4.5	10.5	5	4.5	5
	Distance	44.5	47	61	87.5	59.3
3	Length, cm	22.5	24	19	[15.5]	20
	Width, cm	4.5	7	5	5.5	5
	Distance		53.5	50		
4	Length, cm		20	16.5		
	Width, cm		5.5	4.5		

Numbering of channels starts from the closest to the keel (stern channel in case of support D1C). Figures in square brackets mean that the channel was not preserved completely. Distance was measured between the centres of the mortises.

Reuse of timber was a characteristic of ancient Egyptian boatbuilding.[7] The state of the supports, however, was exactly the same as of the planks of the hull and they bore no evidence of defects or hard wear. At the same time, it seems unlikely that the pieces were prefabricated and taken from a shipyard's stock. The centre-to-centre distance between the channels differs considerably, as do the lengths of the planks, and thus it is difficult to imagine a hull assembled with prefabricated planks of such diverging dimensions. It is possible, however, that the planks were dismantled from an incomplete hull, the construction of which was suspended or cancelled for some reason. Alternatively, these could be pieces from abandoned craft without traces of hard wear. Such traces of wear were also not registered elsewhere in the construction of ship 17.

It was noted above that the supports appear to distribute the weight of the through-beams across the surface of the planking by means of stanchions installed in their upper mortises. The weight of the through-beams would not have been considerable on their own, but it would have been multiplied by the strain originating in the tension of the hull in its transversal axis being held together by these beams. One of the stanchions was discovered in its original position inside bracing timber F9P;[8] it was round and only 7.5 cm in diameter. The stanchions installed inside supports, however, were rectangular in section and, judging by their preserved remains inside upper mortises, more massive. It is worth noting that supports occupy the central area of the hull with the mast step located exactly in the centre of an imaginary square formed by them. It seems that when ship was under sail, this part of the hull was subject to elevated stress and thus needed reinforcement.

5 See this volume, Chapter 9.
6 Channels within supports were 20.5 × 5.7 cm on average against 19.9 × 5.9 cm of those of the planking. The average distance between the midpoints of the channels within supports was 55.6, compared to 46.6 cm in the planking.
7 Creasman 2013; Ward 2000: 95; 2004: 22.
8 See below, 'Stanchions'.

Figure 6.18 Vertical tenon securing the support F8S to the planking (marked by an arrow). (Photo: C. Gerigk; © F. Goddio/IEASM.)

The bracing timbers were installed in specific areas that required reinforcement in the judgement of the shipbuilder.[9] It is possible that they were added one by one in the most vulnerable areas after the ship was already in use.[10] Although bracing timbers reached considerable length (3.28 m in the case of partially preserved timber F1P), they remained 'passive' in their function.[11] With two exceptions, the bracing timbers were not connected to the keel. Bracing timber F6P was secured on the port side by a mortise in the central segment of the keel (K6) and a starboard mortise of same dimensions suggested that there was once another bracing timber installed on the side as well (Figure 6.19). Here again it may be noted that it was the central position of this pair of bracing timbers in the hull that may have made the constructors choose a more solid assembly.

Figure 6.19 Starboard mortise of keel segment K6 viewed from above. Note vegetal remains within mortise that probably sealed the joint. (Photo: A. Belov; © F. Goddio/IEASM.)

Herodotus[12] reports that no ribs were used in the construction of the *baris*; however, it may be that Herodotus saw the construction of a smaller *baris*.[13] Thus it may be that ribs were only used in larger ones. Boreux argued that frameless vessels existed in earlier periods of Egyptian history or, following Pomey, that frames were only added in places they were needed once the ship was in use.[14]

9 The same idea is present in the construction of the Nile *nuggar* (Clarke 1920: 50) and the wadi Halfa boats (Hornell 1942: 34–5).
10 I would like to thank Dr Patrice Pomey for suggesting this hypothesis during the defense of my PhD thesis 31.01.14.
11 The term 'passive' frame, introduced by Basch, means that these frames do not determine the form of the hull and that they are installed once the shell is finished. See Basch 1972.
12 Herodotus *Historia* 2.96, line 8.
13 Boreux 1925: 237.
14 The Dahshur boats do not have frames in their construction. Ward 2004: 20.

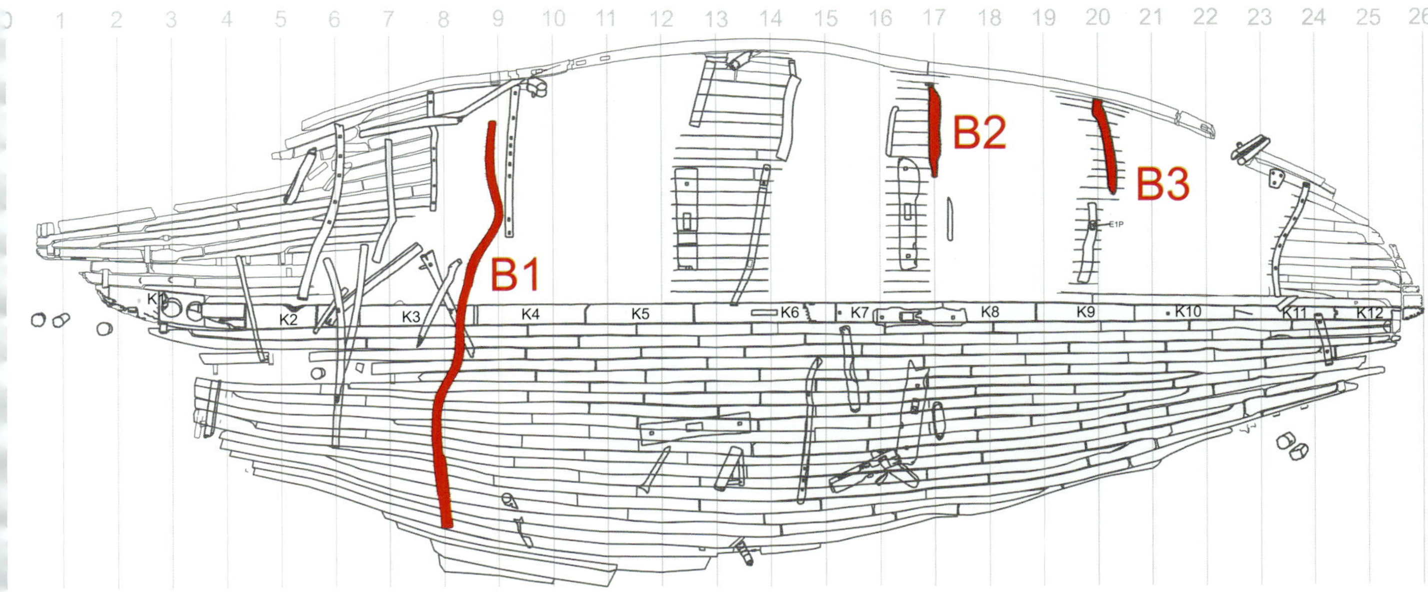

Figure 6.20 Position of the preserved through-beams. (Drawing: P. Sandrin/A. Belov; © F. Goddio/IEASM.)

The Egyptian term *wg* (*wgw*; *wgiw*; *wgyt?*; *wgit?*) probably means 'frames' (*wgꜣ* – rib).[15] The term is found in the *Book of the Dead*[16] and is also attested in Demotic[17] and Coptic[18] sources where the length of the pieces in question could reach 7.5–9 cubits.[19] Archaeological material definitively proves that the Egyptians used frames in the construction of their ships. The Khufu I has 16 frames inserted in the hull with a spacing of 2–3 m between them. Their function, however, was not to support the hull, but to distribute the weight of the deck and of the upper structures on the planking through stanchions.[20] Floor-timbers are visible within the hulls of solar barks cut in the rocks near the pyramids of Giza.[21] The frames from Lisht (1950 BC) are massive and it seems that they formed a real 'bulkhead' supporting the deck.[22] It is also possible that frames were present in the construction of the Mataria boat (*c.* 450 BC).[23] The continued use of frames in Egyptian river craft is also apparent in the ethnographic record.[24]

The structural difference between flat-bottomed and round-bottomed craft has been analysed by Basch,[25] based upon Bell[26] and Marx,[27] who concludes that according to the principle of the 'inverted arch',[28] the hull of round-bottomed craft is strengthened by water pressure. This is not the case, however, for a hull belonging to flat-bottomed craft, which needs the reinforcement of its inner structure. Evidence from ship 17 confirms that the flat-bottomed hulls of larger Egyptian ships, and especially those of freighters, probably required additional reinforcement.

Through-beams

The remains of three through-beams were preserved in ship 17 (Figure 6.20).

Position

The best-preserved through-beam B1 was found in stern region of the ship. Its starboard end still passes through strake S22 (protruding outboard for about 15 cm). The port end of B1 is situated about 70 cm inboard from an

15 The translation of *wg* as 'frames' was initially proposed by Brugsch (1868). For references to Egyptian sources see Jones 1988: 160–61 notes 40–41; *Thesaurus Linguae Aegyptiae*. http://aaew.bbaw.de/. Access Date 06.05.2018: http://aaew.bbaw.de/tla/servlet/BwlBrowser?lcd=wg&tcd=&scd=&pno=1&etr=0&db=0&f=0&l=0&off=14&bc.x=40&bc.y=13&csz=14. Among the research that adheres to this reading are Boreux 1925: 286–93; Junker 1934–1941: 4.74; Hornell 1943a: 27–41; Landström 1970: 20–1, 26, 28, 72; Goyon 1971: 11–41. Goyon suggests the existence of the same 'internal members' in the construction of ships depicted on reliefs of Unas (2321–2306 BC) as those of Khufu I ship. While his meaning is first of all the same piece as the internal girder of Khufu I, it is necessary to note that the pressure from the girder was distributed on the hull by stanchions and frames; Vinson 1994: 22–3, 25.

16 Chapter 19. Boreux 1925: 290–91.

17 Erichsen 1954: 102 note 7.

18 Spiegelberg 1921: 288.

19 Jones 1988: 161 note 40.

20 Ward 2004: 20.

21 Basch 1972.

22 Haldane 1996.

23 Ibid. 1993: 246.

24 Clarke (1920: 50) refers to the construction of Nile *nuggar* in following terms: 'In these days the carpenter occasionally fortifies the hull by a few ribs, but these are in no way parts of a system attached to the keel, but are fixed to the interior of the skin, giving a little extra strength where the builder thinks it desirable.'

25 Basch 1972.

26 Bell 1933.

27 Marx 1947.

28 This principle is well illustrated by the Dahshur boats. The section of planks is complex to allow adhesion between them under the pressure of the water. See Creasman 2010: 120–2.

Figure 6.21 Starboard part of beam B1. (Photo: C. Gerigk; © F. Goddio/IEASM.)

empty mortise in strake P22. No other mortise that could serve for securing the port extremity of the beam was found in this region. It is interesting to note that there is a difference of 48.3 cm between the vertical positions of the two mortises. Even taking into consideration the different deposition of the two planks, it is evident that beam B1 was not horizontal.

Fragments of two more through-beams were preserved at a distance from 17 to 20 m along the baseline on the port side. The preserved ends of these through-beams went through the planking and protruded outboard for some 1.5–2 cm. The three-dimensional position of the through-beams was measured with a goniometre.[29]

Figure 6.22 Through-beam B3 secured in port strakes of ship 17. (Photo: A. Belov; © F. Goddio/IEASM.)

Shape and dimensions

Through-beam B1

The form of the beam was characterised by the natural curvature of the large branch of acacia from which it was made. While the length of the beam reached 7.57 m, the straight-line distance between its ends was only 7.16 m (Figure 6.21).

The flattened starboard extremity of the beam was considerably more massive than that of the port and likely corresponds to the trunk part of the branch. The part of the beam from 1.3–2.5 m from its starboard end can be characterised as being trapeziform in section with the upper side being 8 cm, lower side 17 cm and side faces 12 cm. Towards its port extremity the beam gradually becomes egg-shaped. The width of the beam tapered from 18–21 cm at its starboard extremity to 10.5 cm at its port extremity while its thickness reduced from 12–5 cm.

Through-beam B2

Located on the port side of ship 17, the preserved port beam B2 is 1.85 m long. It was secured in a mortise cut in strake P20. The preserved part of the beam was of rectangular section. Its port end (10 × 5 cm in section) was smaller than that of its starboard (22 × 10 cm).

Through-beam B3

Through-beam B3 was preserved only on the port side of the ship (Figure 6.22), where it was secured in strake P20. Its preserved length is 1.77 m. The average dimensions of this beam of quasi-rectangular section were 6.5 × 4.5 cm.

29 See Figs 11.5 and 11.6.

Figure 6.23 Starboard extremity of through-beam B1 protruding outside the hull, viewed from above. (Photo: C. Gerigk; © F. Goddio/IEASM.)

Figure 6.24 Mortise of strake P22 destined to accommodate the port extremity of through-beam B1, viewed from inside the hull. (Photo: A. Belov; © F. Goddio/IEASM.)

Figure 6.25 Mortise for the port extremity of B1 (strake P22) viewed from outside the hull. Note the tenon for planking assemblage visible on the left inside the mortise. (Photo: A. Belov; © F. Goddio/IEASM.)

Fastenings

The starboard end of through-beam B1 was secured inside strake S22 and it protruded outboard for about 15 cm (Figure 6.23).[30] The traces of a vegetal sealing were discovered within the joint. The port extremity of B1 was initially secured in the mortise in strake P22 measuring 10 × 15 cm (Figures 6.24–6.25). It should be noted that the preserved port extremity of B1 was considerably smaller than the mortise (10.5 × 5 cm) and so it is possible that part of the beam was not preserved. Upon discovery, the port end of through-beam B1 was not in its original position and had become displaced during the decay of the ship. The horizontal shift between the axis of the mortise and the port end of the beam was 24 cm and in the vertical axis the upper surface of the beam was 13 cm above the upper side of the mortise. Through-beam B1 was locked inside the mortise by a planking tenon-rib (Figure 6.26). A small notch, 5 × 3 cm, was cut in the outer lower side of the mortise, which was probably destined for a wedge that could consolidate the joint.

Through-beam B2 was secured to the planking with a small square tenon with a side of 5 cm driven vertically in the mortise of strake P20 (Figure 6.27). Given the

30 Judging by the other two through-beams with their port extremities still in their mortises, it may be noted that B1 projects quite far beyond the outer planking. This is probably a result of post-depositional processes, during which the bulwark of ship 17 was folded inside the hull, thus forcing the beam end to protrude more. I express my gratitude to Dr Damian Robinson for this remark.

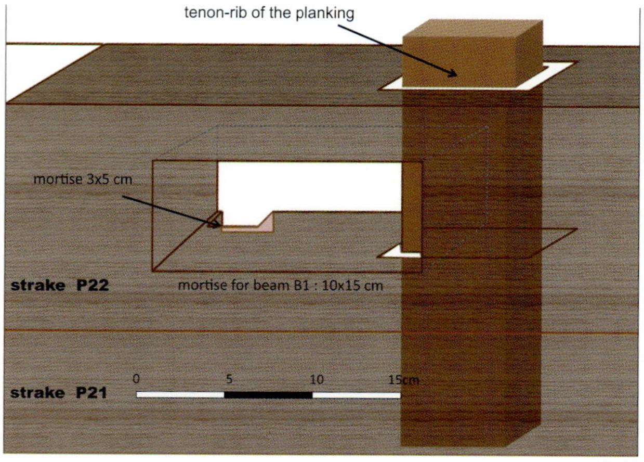

Figure 6.26 Diagram of the locking of beam B1 in the mortise of the port strake P22 viewed from the inside of the hull. (Drawing: A Belov; © F. Goddio/IEASM.)

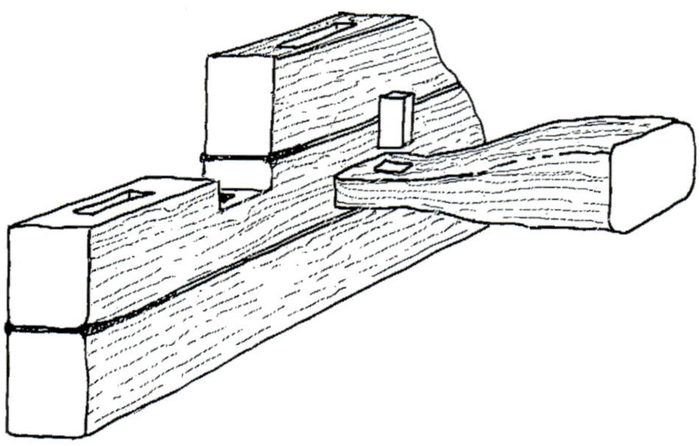

Figure 6.27 Diagram of the locking of beam B2 to the planking. (Drawing: P. Sandrin; © F. Goddio/IEASM.)

Figure 6.28 Through-beam B3 secured in the strake P20. (Photo: C. Gerigk; © F. Goddio/IEASM.)

tapering form of the port end of B2, it may be that it was installed after the strakes P20–P21 were in place.

The joint of through-beam B3 with the planking was not disassembled but the position of the tenon-ribs of the planking seem to confirm that it was secured in the same manner as B1 (Figure 6.28). Like the starboard extremity of B1, that of B3 was also rounded probably to facilitate the insertion of the through-beam inside its mortise. Here again, there were the remains of vegetal sealing in the joint.

Stanchions

Several elements of construction can be classified as stanchions,[31] either on the basis of their position *in situ*, or due to their shape. The remains of stanchion E1 was found in a vertical position inside the mortise cut in bracing timber F9P (Figure 6.29). It was preserved to a height of 39 cm. It was round in section with a diameter of 7.5 cm. The piece was generally not worked and only its end was sharpened to fit into the mortise of F9P (Figure 6.30).

31 The execution of these elements is somewhat rough and alternatively they may be called 'crutches' (cf. 'crutch or stanchion T44 from Mersa/wadi Gawasis', Ward and Zazzaro 2010: 34).

Figure 6.29 Stanchion E1 (marked P1P on the photo) *in situ*. (Photo: C. Gerigk; © F. Goddio/IEASM.)

Figure 6.30 Four sides of stanchion E1. (Photo: A. Belov; © F. Goddio/IEASM.)

Figure 6.31 Construction detail D4S, probably a stanchion, *in situ*. (Photo: C. Gerigk; © F. Goddio/IEASM.)

Figure 6.32 Concave end of detail D4S. (Photo: A. Belov; © F. Goddio/IEASM.)

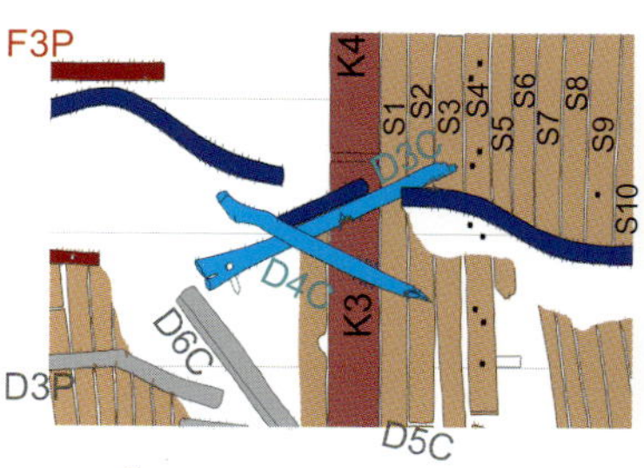

a

b

Figure 6.33 Relative position of the through-beam B1 and construction details D3C and D4C shown in (a) plan and (b) photograph. (Drawing: P. Sandrin/A. Belov; photo: C. Gerigk; © F. Goddio/IEASM.)

Constructional detail D4S was found inside the hull on the starboard side of the ship (Figure 6.31).[32] Oval in section, it is apparently intact: 96 cm long, 10 cm wide and from 5–6 cm thick. One of its ends was sharpened in the same way as that of stanchion E1. The other end of D1S, considerably larger, was deltoid in shape, its upper surface worked with an adze to form a perfect concave curve (Figure 6.32).

Constructional detail D4C was intact and 1.74 m long. It was found lying parallel to through-beam B1 (Figure 6.33). While its north end was sharpened in the same way as E1 and D4S, the southern one was cut square. The results of the reconstruction of the hull suggest that this piece could be a stanchion. According to the model discussed in Chapter 11, the distance between the upper surface of the keel and through-beam B1 in this area was very close to the length of D4C.

The function of constructional detail D3C as a stanchion is more difficult to prove. This piece, found under through-beam B1, was preserved for 2.05 m but was probably not complete (see Figure 6.33). D3C had a tenon passing through its deltoid southern end that resembled that of D4S. The piece was too tall to serve as a vertical stanchion in this area; however, it could have been employed as an inclined stanchion.

32 X = 1.20 m, Y = 2.42 m.

Conclusions

Through-beams were a characteristic feature of Egyptian boatbuilding[33] and are clearly seen in the iconographic record. The earliest example is seen in the model of a boat of Prince Mesehti from the First Intermediate Period (*c.* 2160–2055 BC).[34] In representations ships can contain several levels of through-beams. Three levels of beams are visible on the images of the transport barges for the obelisks of Queen Hatshepsut (1473–1458 BC).[35] Another example, which is also the latest known for this feature, is that of 'the Great Ship of Saïs' on the so-called 'blocks of Piankhy' (Psamtik I, 664–610 BC).[36] The ship in question has two levels of through-beams attesting to her large dimensions (see Figure 7.6).

In ancient Egyptian boatbuilding, through-beams were an essential feature of construction and had several distinct functions. First of all, they strengthened the hull in the transversal axis.[37] Secondly, through-beams were an efficient means to strengthen the *longitudinal* structure of ship as well.[38] Thirdly, the through-beams distributed the weight of the deck and any cargo (which was often placed above it) over the entire surface of the hull.[39] Finally, the beams served as thwarts for rowers.[40]

Contrary to the widespread boatbuilding method of placing beams on the top timbers of the opposite branches of the ribs, the ancient Egyptians incorporated the through-beams into the strong planking.[41] Archaeologically, this phenomenon is attested in the Dahshur boats[42] and by beam T34 from Mersa/wadi Gawasis.[43] Herodotus mentions beams in his description of the *baris*.[44] The construction of the traditional river-faring boat of Sudan, the *nuggar*, includes through-beams, which are incorporated into the planking.[45]

The four massive supports that occupied the central section of ship 17, and contained the remains of stanchions, served to distribute the strain from the beams onto the larger surface of the hull. Ten preserved bracing timbers consolidated the hull in selected areas. With two exceptions they were not attached to the keel. It seems that the through-beams were aligned above the supports and some of bracing timbers. According to available iconographic and archaeological material the through-beams of ancient Egyptian ships were regularly distributed along the hull. Even if the beams of ship 17 had a rather asymmetric shape and were coarsely worked, they must have followed a regular distribution pattern. It seems that not all beams were preserved and it is likely that a middle beam also existed to support the mast.[46] Three through-beams were preserved, but at least seven can be hypothesised.[47] The through-beams of ship 17 were locked within the channels of the 20th–22nd strake by the tenons of the planking to give more solidity to the hull in the transversal axis.[48]

For the construction of beam B1, the shipwrights chose an irregularly-shaped branch of acacia. The trapezoidal section of beam B1 and the tapering extremities of beams B2 and B3 find parallels in the material from Mersa/wadi Gawasis.[49] The absence of any traces on the upper surface of beam B1, and its inclined position within the hull, testify that ship 17 was undecked. The roughness and strength of the beams of ship 17 would indicate that they belonged to a cargo boat.

In ancient Egypt, stanchions[50] are seen in the construction of the Khufu I ship (*c.* 2500 BC),[51] the Dahshur boats (*c.* 1878–1841 BC)[52] and by a crutch or stanchion T44 from Mersa/wadi Gawasis (probably Middle Kingdom).[53] The concave upper extremity of stanchions D4S and D3C was probably destined to support the central part of beam B1, which was rounded in section. This would explain the absence of any mortises on the lower surface of the beam.

33 The technical solution of through-beams can be found in some other boatbuilding traditions of the world: see Pâris 1841: 15, pl. 8, figs 1–3; 49, pls 46–7.

34 See Landström 1970: fig. 207.

35 See Bell 1934; Landström 1970: 128–33; Sleeswyk 1987; Sölver 1943.

36 See Perdu 2004: 226.

37 Jones 1995: 76. 'These beams not only supported the weight of the deck-planking, but also served to provide athwartship support to the hull and prevent the sides from sagging outwards'.

38 It seems that this function of the through-beams has not been made the subject of a detailed study. I am very grateful to Juan-Pablo Olaberria for suggesting several simple physical experiments that visually prove this conclusion.

39 Jones 1990: 54.

40 Boreux 1925: 332.

41 Faulkner 1940: 4.

42 Ibid.

43 Ward and Zazzaro 2010: 32.

44 Chapter 10.

45 Clarke 1920: 49; Hornell 1943a: 29.

46 Cf. model of Prince Mesehti, First Intermediate Period (*c.* 2160–2055 BC). Note that the middle through-beam is larger than the others (see Landström 1970: 210); model from the tomb of Karenen in Saqqara (First Intermediate Period, see Quibell 1908: pl. 26); models from the tomb of Meketre (Middle Kingdom, Eleventh dynasty, *c.* 2055–1985 BC, see Winlock 1955: 45–69). Although the models dating to the New Kingdom are not so informative in this aspect because of the large deck-house that hides the base of the mast, it was probably always supported by a middle beam. Cf. model of Amenhotep II (1427–1400 BC, see Landström 1970: 108–9); boat on the relief from the tomb of Sennefer (*c.* 1439–1413 BC, see Edgerton 1922–3: 125, fig. 9); boat on the relief from the tomb of Rekhmire (eighteenth dynasty, vizier under Thutmose III and Amenhotep II, see de Garis Davies 1935: pl. 61); ships of the Punt expedition of Queen Hatshepsut (1473–1458 BC) and warships of Ramesses III (1184–1153 BC), see graphical reconstructions of Landström 1970: 108–9, 113, 125–6, 136–7, 139. The middle beam of the *nuggar* is 'usually made from the stem of a tree, smoothed with the adze, but following all the inequalitites of its shape' (Clarke 1920: 49).

47 This estimate is based on the distance of three meters between the beams B2 and B3.

48 In his analysis of models from the tomb of Amenhotep II (1427–1400 BC), Landström (1970: 109) suggested that 'The beams were probably locked between the planks, thus giving the vessel great strength athwartships'.

49 Beam T34. See Ward and Zazzaro 2010: 32.

50 *ips*, see Jones 1988: 153, note 7; *isp.t* (?), see Glanville 1932: 15–6.

51 Lipke 1984: 119, note 11, figs 32, 74.

52 Ward 2000: 87.

53 Ward and 2010: 34. Cave 4 which contained the stanchion in question was not precisely dated. It is interesting to note that the length of this piece (1.89 m) is close to the length of stanchion D4C of ship 17 (1.74 m).

7 The steering system

Figure 7.1 Aftermost segment K1 of the keel of the ship featuring two circular openings. (Photo: C. Gerigk; © F. Goddio/IEASM.)

Archaeological remains

The aftermost segment (K1) of the keel of ship 17 was a single piece of acacia wood with the following dimensions: length 3.05 m, width 56 cm, with a thickness that varies from 10–21 cm.[1] The detail was generally well-preserved, except for the eroded starboard stern end. In addition there were two transversal and two longitudinal cracks, the result of post-depositional deformation of the hull (Figure 7.1). Segment K1 had an inclination of 43.5° to the port.

In plan, segment K1 consisted of a rectangular part 1.66 × 0.56 m and a trapezoidal part tapering to a width of only 15 cm to the stern. Two circular openings with diameters of 33 and 39 cm were carefully cut with a small chisel in the centre of segment K1. The openings go through the entire thickness of the keel. The first opening was damaged, but the detached pieces of wood were found in their initial position. The shortest distance between the edges of the openings measured only 50 mm. Both openings were almost vertical, both having an angle of 92° in reference to the horizontal axis of segment K1. In profile segment K1 was characterised by the greater thickness of the rectangular central part, which corresponds to the position of the openings (Figure 7.2). Segment K1 was joined to the neighbouring segment K2 by means of the same tongue-and-groove joint that was used for the assembly of all the other parts of the keel. Six tenons passed through segment K1 and

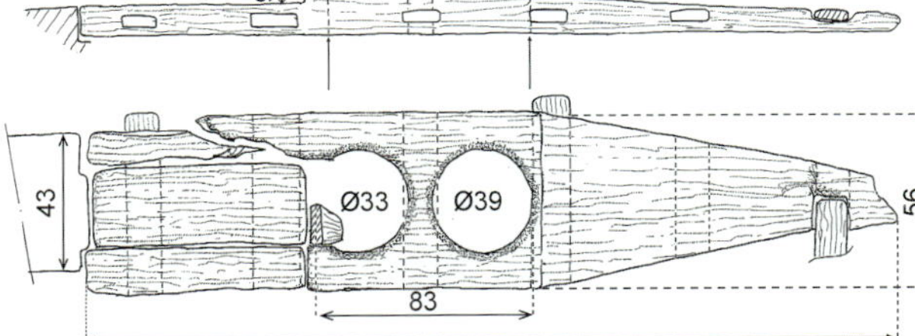

Figure 7.2 Plan and profile of the segment K1 featuring two circular openings. (Drawing: P. Sandrin; © F. Goddio/IEASM.)

1 Belov 2014c.

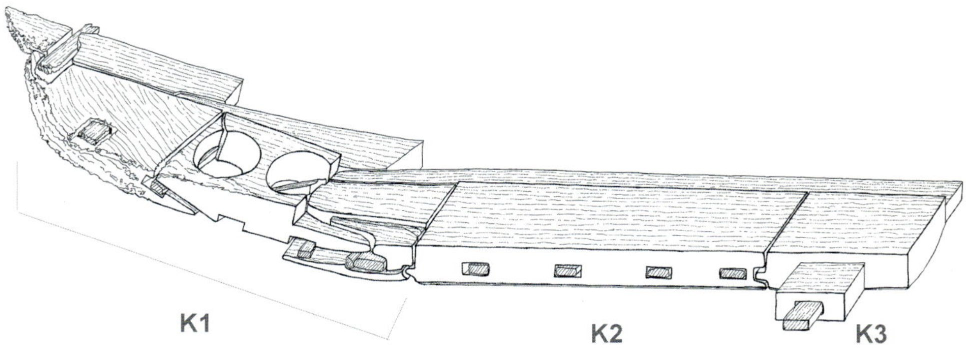

Figure 7.3 Axonometric view of the stern segments K1, K2 and K3 of the keel. (Drawing: P. Sandrin; © F. Goddio/ IEASM.)

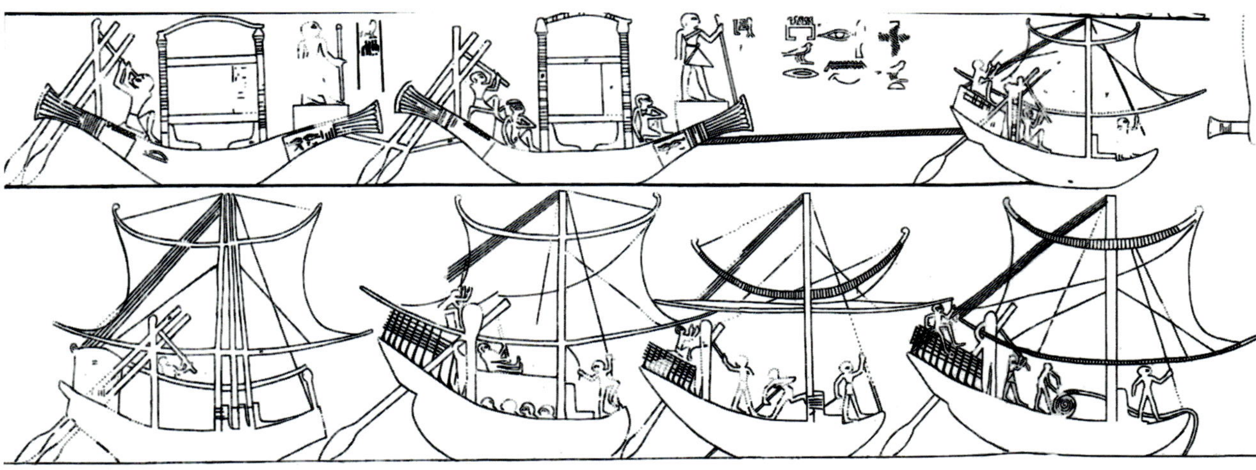

Figure 7.4 Boats from the reliefs at Deir el-Gebrâwi, sixth dynasty (c. 2325–2155 BC). (Drawing: A. Belov; after de Garis Davies 1902: pl. 6.)

secured it to the strakes of both boards (Figure 7.3). These tenons passed through the keel only in places where they did not obstruct the openings. The unpreserved tenon between the openings was most probably trimmed to leave the apertures free.

Discussion

The first representations of the axial rudder in Egypt are dated to the end of the fifth (*c.* 2494–2345 BC)[2] or to the sixth dynasty (*c.* 2345–2181 BC). The best-known are the sixth dynasty reliefs from Deir el-Gebrâwi, Saqqara.[3] The axial rudder and the alternative arrangement of a pair of quarter-rudders present on these reliefs continued to co-exist throughout Egyptian history; seemingly, the axial rudder was the preferred steering system for Nile freighters and travelling ships (Figure 7.4).[4]

The development of the axial rudder continued in the First Intermediate Period (*c.* 2160–2055 BC) and in the Middle Kingdom (*c.* 2055–1650 BC). Hundreds of models of passenger or travelling boats dated to the Middle Kingdom were equipped with an axial rudder (Figure 7.5). The rudder-stock passed through a cleft at the stern of the boat and was secured to a larger stanchion than was used in earlier periods. In comparison with Old Kingdom examples, the surface of the rudder was also considerably enlarged. In the New Kingdom (*c.* 1550–1069 BC) the cleft at the stern became more pronounced and often, in order to make it visible on the drawing, the artists turned it towards the viewer.[5] The upper extremity of the rudder-stock was sometimes crowned with a massive rectangular piece that probably served as a counter-balance to facilitate the lifting of the rudder from the water.[6]

Figure 7.5 Typical steering arrangement of a Nile boat in the Middle Kingdom. Detail of a model of a boat (AN1896-1908.E.2296) from Beni Hasan (*c.* 1975–1640 BC). (Photo: Ian. R. Cartwright 2014. Image © Ashmolean Museum, University of Oxford.)

2 Jones 1995: 39–40.
3 Davies 1901–1902; Boreux 1925: 402.
4 Boreux 1925: 397.
5 Doyle 1998: 56.
6 Landström 1970: 134.

8 The remains of the rigging

Figure 8.1 Mortise in central segment K6 of the proto-keel of the ship viewed from above. (Photo: C. Gerigk; © F. Goddio/IEASM.)

Very little of ship 17's rigging has survived.[1] Normally, ancient Egyptian freighters were sailing boats and Herodotus' text confirms that the *baris* was not an exception. Images from the Middle Kingdom onwards show the mast of Egyptian ships placed amidships.[2] Consequently, it was not surprising to find an elongated mortise in segment K6 of the keel of ship 17 (Figures 8.1–8.2). The mortise was 46 cm long, 13 cm wide and about 5 cm deep.

There were no other traces on the upper surface of segment K6 and it is possible that the mast spur was installed directly into the mortise, which was probably originally of greater depth as the segment is eroded. To cite an ethnographic parallel, the mast of Sudanese *nuggar* is installed in a socket in the keel.[3]

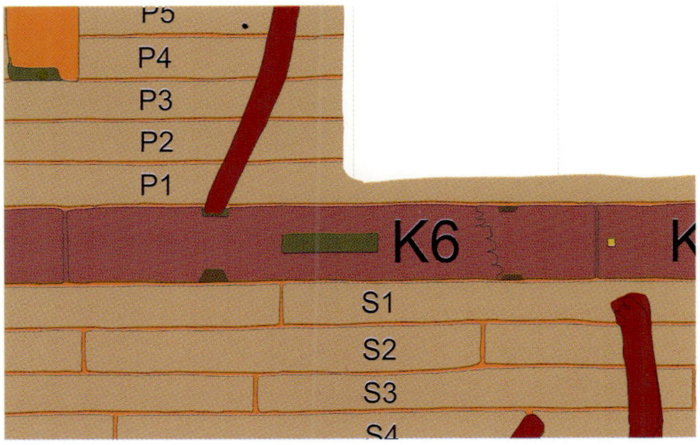

Figure 8.2 Detail of the general plan of the ship showing the segment of the keel K6 with the mortise in its centre. (Drawing: P. Sandrin/A. Belov; © F. Goddio/IEASM.)

1 This is true for the other ships from Thonis-Heracleion as well. Rigging was often reused in antiquity and a mast, for example, had a particular value. See Lewis 1960. Rigging would also have been amongst the first things to be removed from the ship during its primary salvage.

2 Assmann 1913: 148–9.

3 Clarke 1920: 49, 'The stout beam or tree stem is for making steady the short mast which has a socket in the keel and a strap or other form of stay to secure it to the beam'.

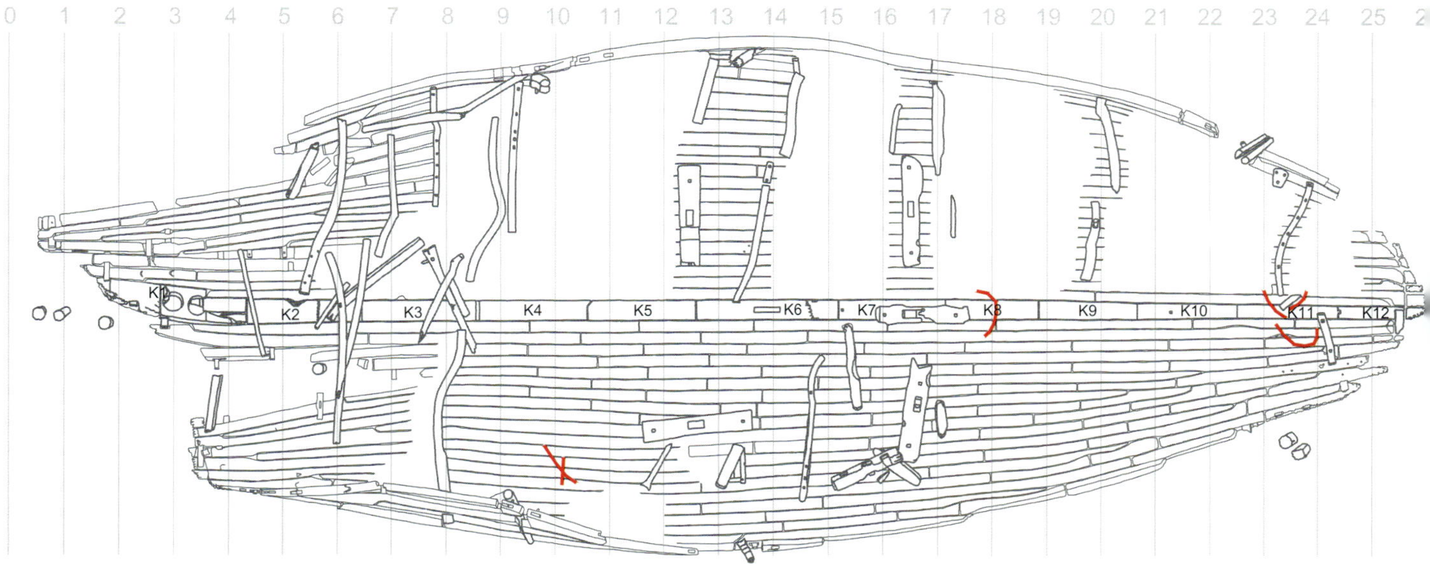

Figure 8.3 Position of fragments of cordage on the general plan of ship 17. (Drawing: P. Sandrin/A. Belov; © F. Goddio/IEASM.)

Figure 8.4 Fragment of cordage found on the starboard at 10 m along the baseline (10, C). (Photo: A. Belov; © F Goddio/IEASM.)

The mast of ancient Egyptian vessels was usually removable and although centrally-positioned, its easy removal was obstructed by the deckhouse of New Kingdom ships,[4] although it was possible. An estimate of the height of the mast can be made on the basis of a general ratio to the length of the hull that was close to 2:3 for ancient Egyptian ships.[5] For ship 17 this would give a mast *c.* 17–18 m high.

Several fragments of cordage were found during excavations inside the hull of ship 17 (Figure 8.3). Their length did not exceed one metre and they were not connected to elements of construction.[6] It appears that all of the discovered ropes belonged to the same type with a diameter that varied from 50–60 mm. The ropes were composed of two 'Z'-twisted strands, each consisting of three yarns (Figures 8.4–8.5).[7]

As a comparison, the ropes from the Khufu I ship were of three different sizes: 5, 8 and 13 mm.[8] Cordage from Mersa/wadi Gawasis were 30–35 mm in diameter.[9] Therefore it seems that the ropes used on ship 17 were

4 Jones 1995: 49.
5 Goyon 1971: 22.
6 The samples of the material were collected, but are yet to be analysed.
7 For ancient rope-making technology and terminology see Charlton 1996; Sanders 2010; Veldmeijer 2009.
8 Landström 1970: 28.
9 Ward and Zazzaro 2010: 39.

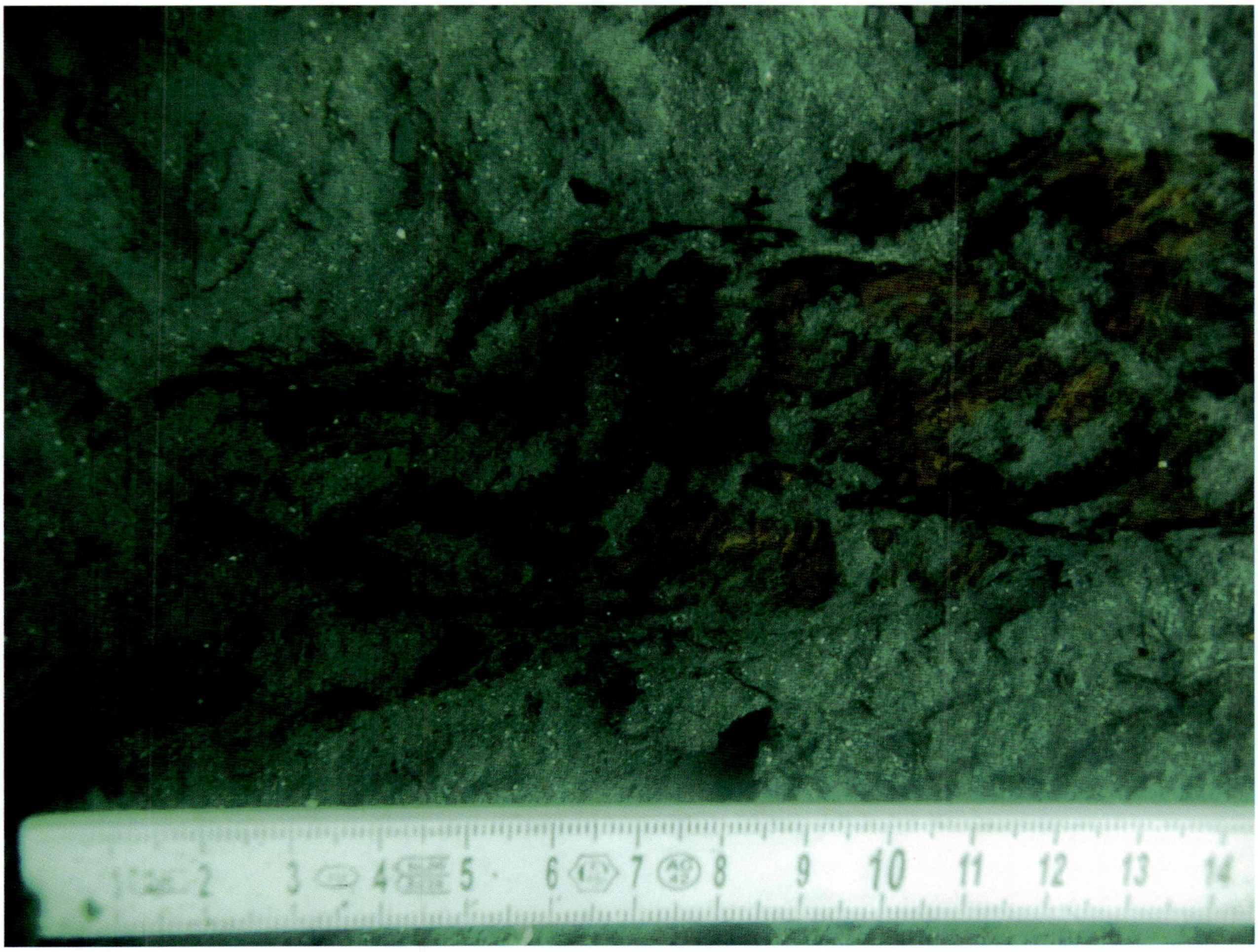

Figure 8.5 A detail of the rope (10, C). (Photo: A. Belov; © F. Goddio/IEASM.)

quite large. This fits within the overall robust character of the ship gained from an examination of the planking and its suggested role as a freighter.

No traces of sail were discovered on ship 17. According to Herodotus the sail of the *baris* was made of papyrus,[10] however, only sails of flax have been attested archaeologically in Egypt.[11]

10 The same application of papyrus is also mentioned in *P.Cair. Zen.* 4.59566 (*c.* 250 BC). See Arnaud 2015a: 109.

11 For sails from Lyon (second half of the third century BC), see Rougé 1987) and from Berenice (first century BC), see Wild and Wild 2001.

III

The construction sequence, vessel type and technical evaluations of the hull

9 The construction sequence[1]

The type of construction seen in Thonis-Heracleion is not attested elsewhere in the archaeological record (perhaps with the exception of the fragmentary data from the Mataria boat) and the only contemporary parallels are provided by Herodotus. The planking of ship 17 provides the strength of the hull, which the through-beams help to tie together. Other bracing timbers and supports are of secondary/lesser importance in helping to stiffen up the hull. The ship was built of local timber and there is evidence for a frugal use of building material.[2] The five massive supports of the ship were built of reused planks, which might be considered another typical feature of Egyptian boatbuilding.[3] The information at present is insufficient to provide a definitive construction sequence for ship 17, so the following observations are preliminary reflections to be refined as further evidence emerges from the other *baris*-type ships in Thonis-Heracleion.

Stage 1

The transversal tenon-rib channels were initially cut in the lower third of the keel's thickness (moulded dimension). Four or five channels were cut into each segment of the keel depending on its length. The channels were distributed in a regular fashion with an average centre-to-centre distance of 47 cm. The shipwrights appear to have cut the two outer channels very close to the extremities of each segment. After that, all of the segments of the keel were assembled. Keeping in mind the short lengths of the keel segments, it seems probable that it was easier to assemble them all lying flat on the ground and then to raise the extremities in order to obtain the desired curvature of the hull, rather than joining them one after another. The segments are assembled with tongue-and-groove joints that are not deep and leave enough play for this.

The keel ended with massive stem and stern timbers that were not preserved in ship 17, but are presented in other vessels of the Late Period from Thonis-Heracleion.[4] It was at this stage that the two shafts for the axial rudder were probably cut in the stern segment of the keel, seen in the specific box-like configuration of segment K1. Thus the boatbuilders started with a clear idea of the draft of the craft under construction and of its final position in the water.

Stage 2

The proto-keel was laid out in the form of a crescent with supports placed under it. Taking into consideration the considerable length and weight of the acacia keel, a device of some kind must have been employed to support the hull during its construction. It could have been a rope truss, as is seen on many reliefs of the Old Kingdom.[5] A rope truss would have been used only during construction and should not to be confused with a hogging truss. Although this element disappears from the iconographic record after the Old Kingdom, Egyptian shipwrights probably continued to use it for larger vessels.[6] As the tongue-and-groove joints between the segments of the keel are not deep, it is possible that the first strake was added to the keel before it was raised into its curved form (see Stage 4a below).[7]

The outer surface of the keel was shaped using adzes. This stage is seen by adze marks on the outer surface of keel segment K6. It is worth noting that the adze marks followed the grain of the wood and thus the outer surface of the keel is rather uneven. The planking was similarly adzed, illustrating that the hull corresponds to a massively constructed, roughly hewn working boat.

Stage 3

A first set of tenon-ribs was inserted into the channels cut into the proto-keel so that the proto-keel lay at the middle of each tenon-rib (Figure 9.1). The ends of the tenon-ribs were slightly rounded using a knife to facilitate the insertion of the planks. Each tenon-rib was secured tightly inside the mortises by four wedges. The garboard (which does not differ morphologically from the other strakes) must have been added at the next stage, once the wedges had been driven into place. It is quite possible that Stage 3 took place simultaneously with Stage 4 and that the tenon-ribs were installed in the keel in groups,

1 Belov 2014b.
2 Cf. Ward 2000: 14 on this point with regards to other Egyptian vessels.
3 Creasman 2013; Ward 2000: 140.
4 See this volume, Chapter 4 and Robinson 2015.
5 Landström 1970: 39; Sliwa 1975: 58.
6 Rogers 1996: 99–104.
7 It is possible, however, that the tenons of these strakes were not wedged within the channel in order to have more flexibility (see below).

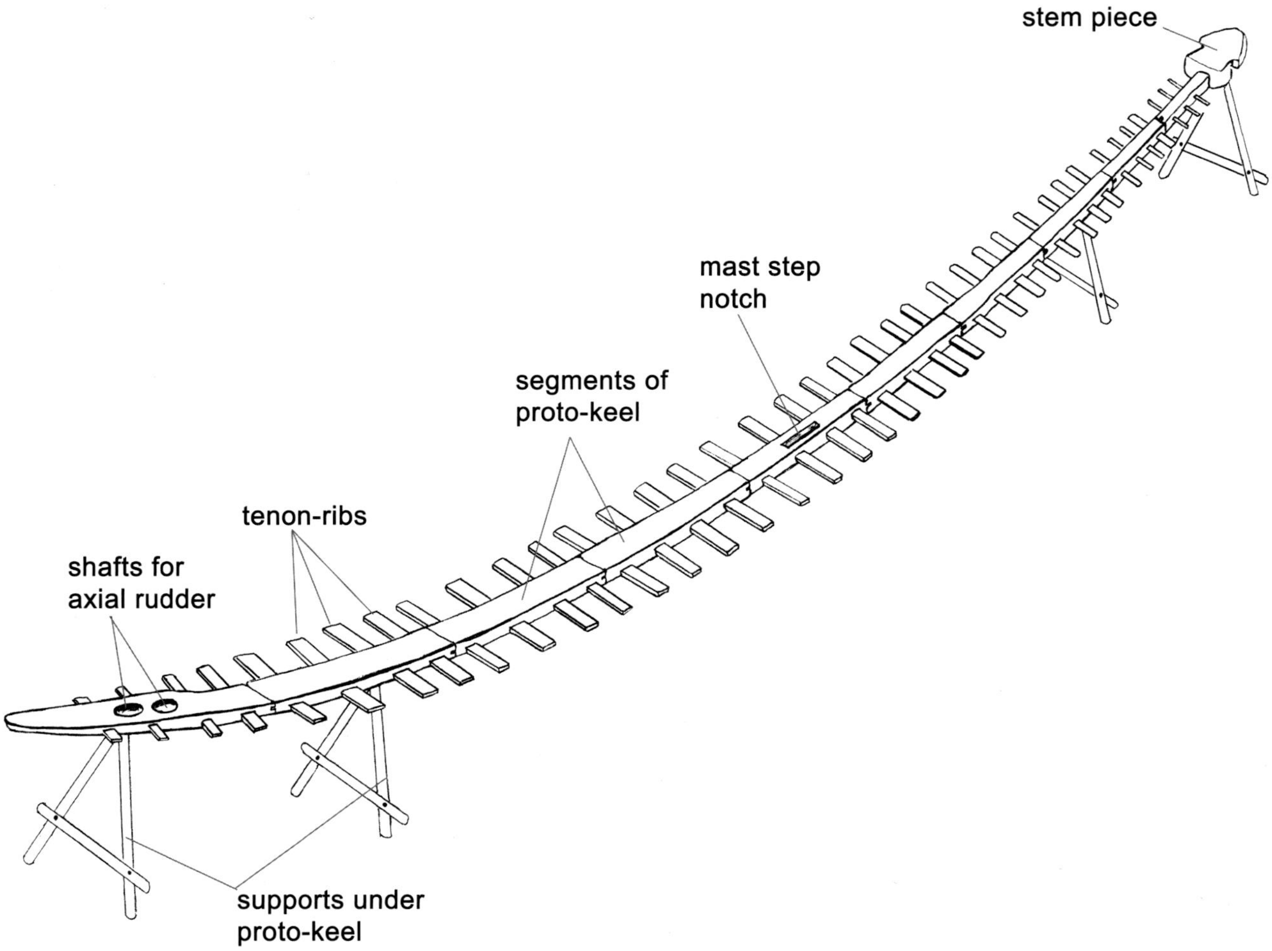

Figure 9.1 Stage 3 in the construction of the ship. Installation of the tenon-ribs in the proto-keel. Not to scale. (Drawing: A. Belov.)

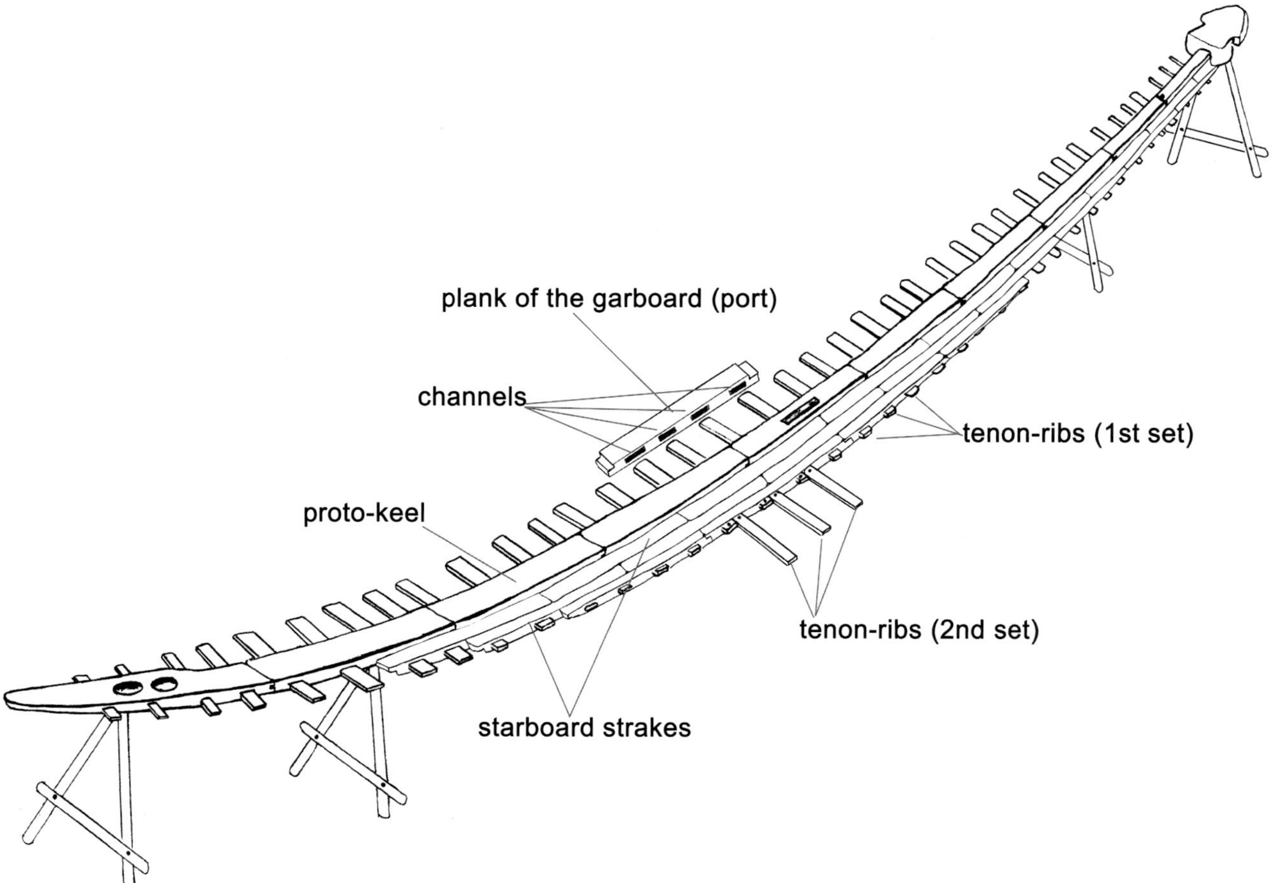

Figure 9.2 Stage 4a in the construction of the ship. Installation of the first four strakes. Not to scale. (Drawing: A. Belov.)

and planks attached gradually starting from the centre of the ship.

Stage 4a

The planks of particular lengths were used to ensure that the joints between the planks and proto-keel would be off-set (Figure 9.2). The extremities of the planks were shaped to form the half-laps (in the majority of cases) used to join planks in a strake. It is probable that the primary adjustment between the planks of the garboard strake and the proto-keel was carried out before the tenon-ribs were inserted. Then the planks of the garboard strake had channels cut to correspond to the position of the tenon-ribs installed in the proto-keel. The installation of the planks probably started from the centre of the ship with planks added one after another, rather than as a complete strake. It would be impracticable to insert an entire strake consisting of short, thick planks assembled by loose half-lap joints while simultaneously aligning the numerous tenon-ribs with their channels. This conclusion is further supported by iconography, including the boatbuilding scene from the tomb of Khnumhotep from Beni Hassan (Middle Kingdom) among others, which suggests that in the Old and Middle Kingdom planks of the Egyptian boats were joined in a 'pyramidal style' moving from the centre of the ship towards the extremities (Figure 9.3).

This style of construction would have appeared very strange to a boatbuilder from outside Egypt, which is perhaps why it was reported by Herodotus in detail. He explicitly comments on 'the strong and long tenons they insert two-cubit planks', which resembles brickwork in the way the planks were assembled (brick after brick).

The rectangular sections of the proto-keel and of the planks of the first four to five strakes are assembled using the same tenon-ribs passing through rectangular channels. This indicates that the bottom of this ship was flat in the transversal axis.

Preliminary adjustment was necessarily followed by the precise fitting of each plank. Simple wooden compasses, like the *sheba* used by the Sudanese boatbuilders, could have been used to help obtain the correct curve along the lower edge of the next plank to be added to obtain tight joints.[8] Such a system of assembly did not allow for the channels to be cut with much accuracy and this might explain the considerable room left for the tenons inside the channels. Thus, the average width of the regular channels of the starboard is 19.1 cm, while the average width of the respective tenons is only 12.8 cm. After the next plank was added, wedges were inserted tightly into the channels, thus maintaining the strength of the planking.

8 Hornell 1943a: 29.

Figure 9.3 Boatbuilding scenes from the tomb of Khnumhotep at Beni Hasan (Middle Kingdom). (Drawing: A. Belov; after Newberry 1893: pl. 29.)

Planks making up the garboard strake were added in this way, working towards the vessel's extremities. It seems that assembling planks to the massive stem and stern pieces was an important stage. The last planks of the strake were probably joined to them after the central planks of the subsequent strakes were already in place. The boatbuilders must have used shorter tenons that were inserted with a different angle than for the rest of the strake. This part of the construction requires further research, as these are the areas of ship 17 that were least well-preserved.

The planks of the first four to five strakes in the bow area were assembled as described above. Luting the joints must have accompanied the assembly of the planking and corresponds to the words of Herodotus: 'They obturate the seams from within with papyrus'.[9]

Stage 4b

At strake four, the tenon-ribs which pass through the keel were pegged and a second series of tenon-ribs was installed and also pegged. For this purpose, a double-sized channel was cut in strake four, having an average width of 31 cm. The tenon-ribs of the first series ended in closed mortises within strake five. Thus adjacent tenon-ribs had a two-strake overlap (S4 and S5). The second series of tenon-ribs assembled the planking from strakes S4–S5 to strakes S10–11 (in general seven to eight strakes). Starting from S5, knife-shaped planks were employed to close the bow and stern areas of the hull.

Stage 4c

At the next stage, the third series of tenon-ribs comes into play, assembling the starboard strakes from S10–11 to S15–16 (Figure 9.4). The upper end of one tenon-rib and the lower end of its neighbour were pegged to the planking. Note that in the bow, and probably also in the stern area, the tenon-ribs could assemble more strakes than in the central part of the ship. Thus, at the bow strakes from S4 to S16 were assembled by a single series of tenon-ribs.

9 See this volume, Chapter 10; Herodotus *Historia* 2.96.

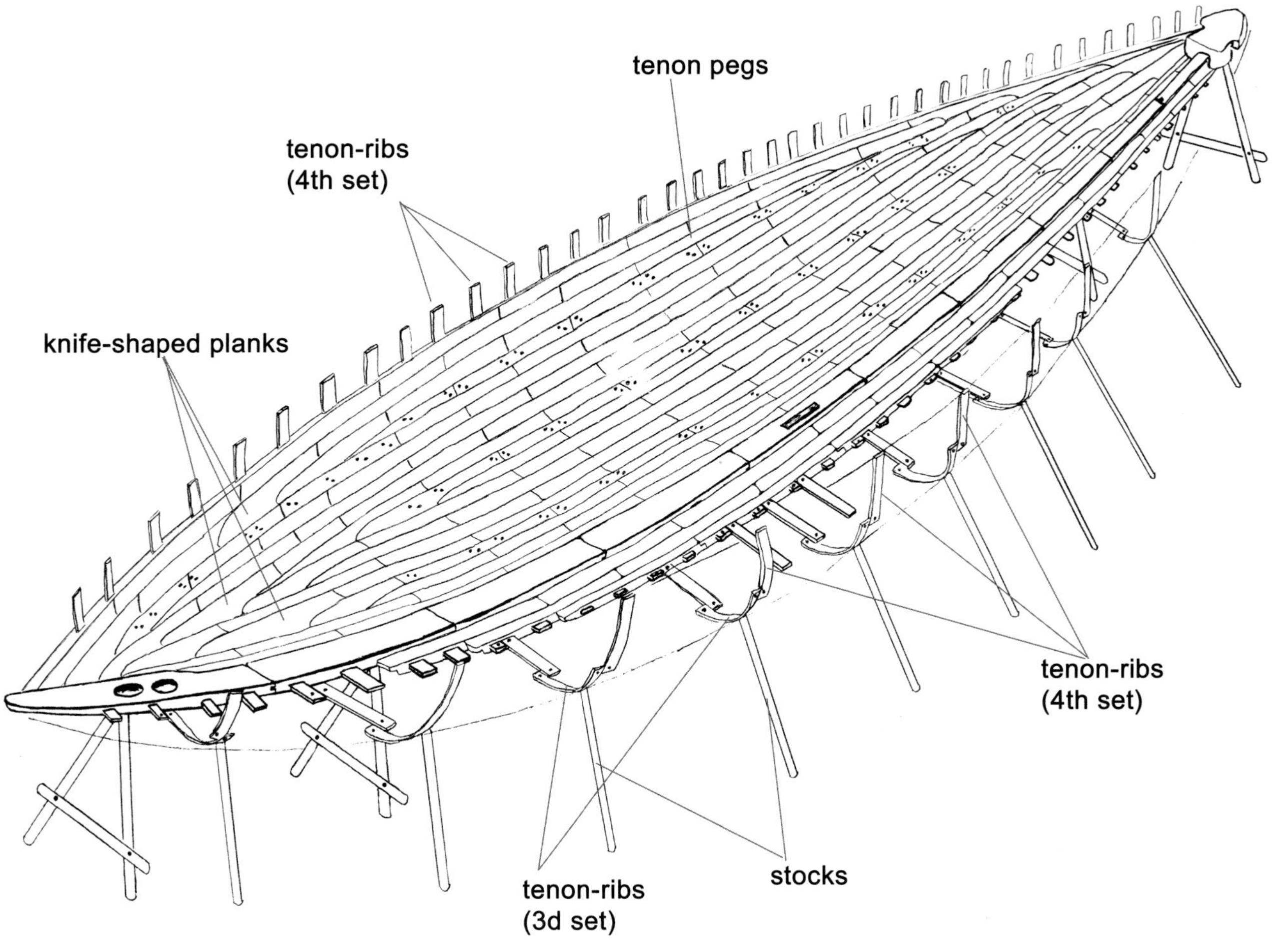

Figure 9.4 Stage 4c in the construction of the ship. Assembling the strakes from S10–11 to S15–6 with the third set of tenon-ribs. Not to scale. (Drawing: A. Belov.)

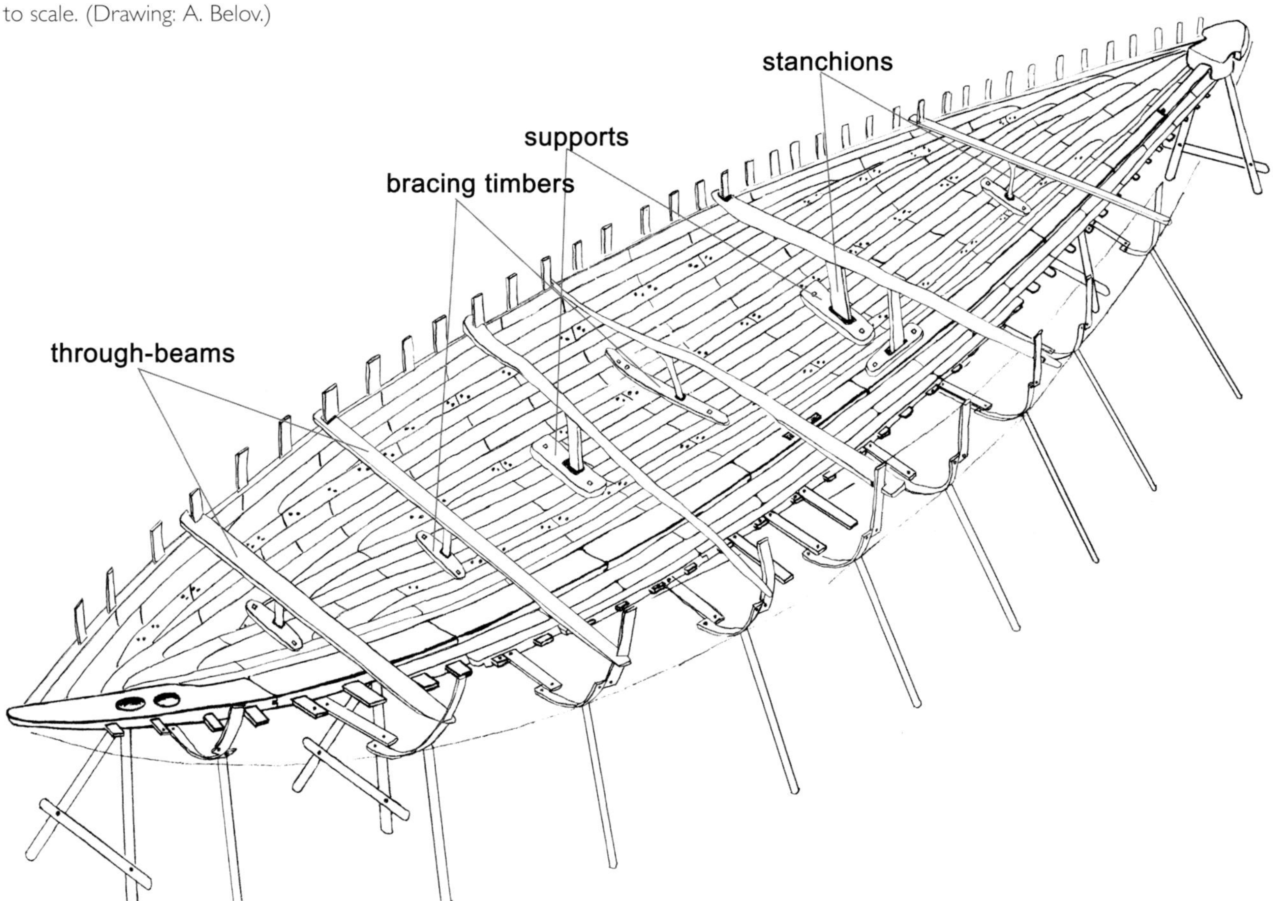

Figure 9.5 Stage 5 in the construction of the ship. Installation of the through-beams. Not to scale. (Drawing: A. Belov.)

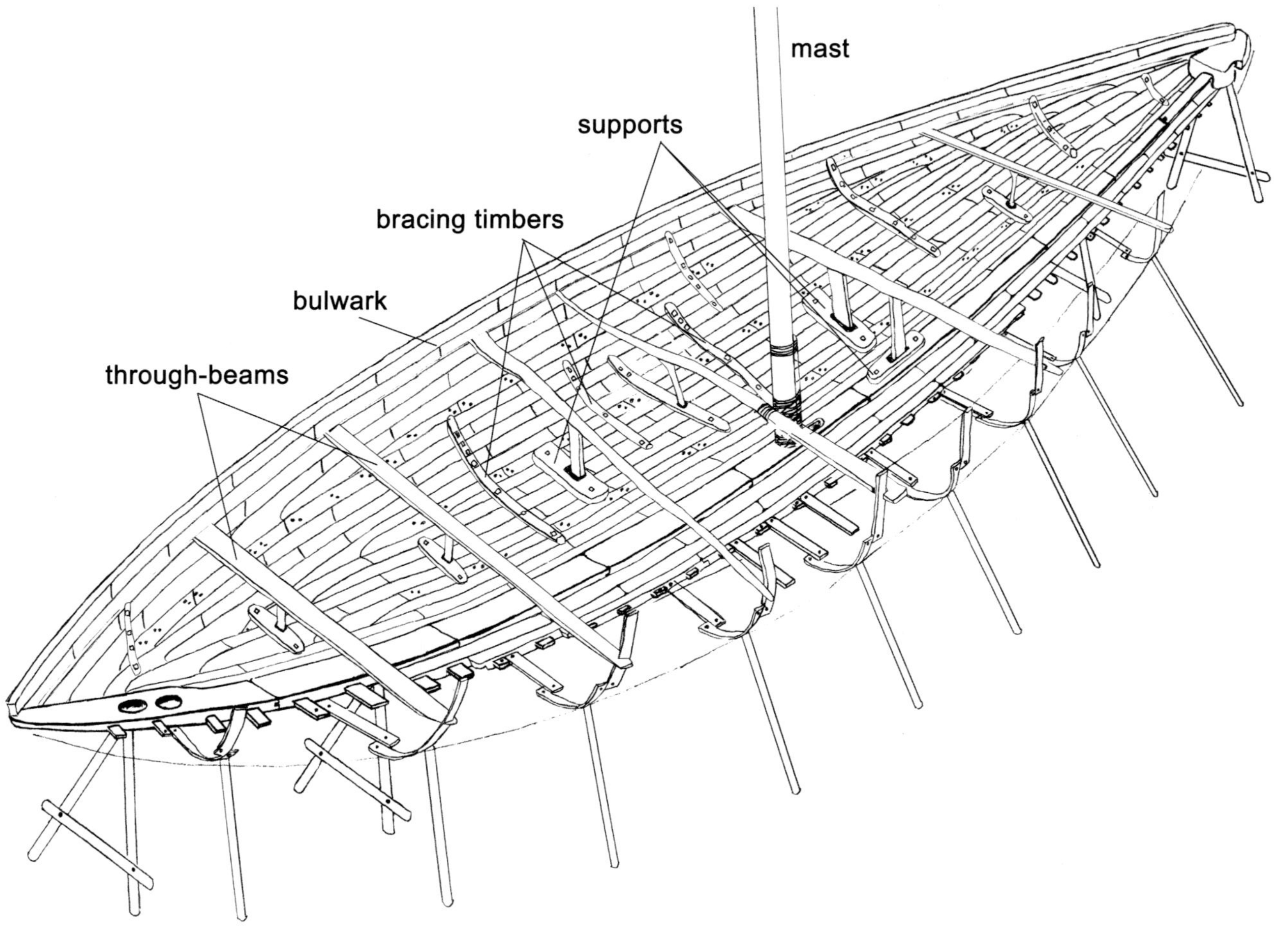

Figure 9.6 Final stages in construction of the ship. Not to scale. (Drawing: A. Belov.)

It seems that the floor remained relatively flat until at least strakes 11–13. Arguments in favour of this conclusion include the length of the completely preserved through-beam B1 (that limits the beam of the ship), the position of four of the massive and inflexible supports made of reused planks between strakes 4 and 12–13, and the distribution of the squared pegs fixing the bracing timbers to the planking. These pegs are placed more densely from starboard strake S13 (or port strake P11) towards the bulwark, probably indicating the increased curvature of this region. At this stage of construction, it was necessary to support the sides of the ship to maintain the desired shape. It seems unlikely that the constructors applied cross-spawls, like those used during the construction of Kyrenia II,[10] because they would be ineffective with short planks assembled one after the other as suggested here. Shipyard stocks supporting the hull from the outside seem more likely.

A few pegs were preserved in the subsequent strakes (S17–S18), suggesting that the fourth series of tenon-ribs assembled the strakes from S15–16 to S24.

Stage 5

Through-beams, made of naturally curved compass timber, were installed at the level of the starboard strake S22 and port strakes P20–P22 (Figure 9.5). The ends of the through-beams were locked into place inside the planking by the fourth series of tenon-ribs. Five massive supports were attached to the planking in the central part of the hull under beam B2P and under two other hypothetical beams at 12 and 14 m from the preserved aft extremity. The latter beam, of which only fragments remain, probably supported the mast. Stanchions were installed in the upper mortises found in the supports and of some of the bracing timbers in order to distribute the weight of the beams onto a larger area of the planking.

The fourth series of tenon-ribs probably assembled the bulwark strakes. The representations of the Old Kingdom show the process of installation of a bulwark as an entire strake, rather than plank-by-plank as for the previous strakes.[11] The poor state of preservation of the upper strakes makes it difficult to judge whether this was the case for ship 17, but it cannot be excluded. The last preserved strakes seem to indicate that the bulwark was quasi-vertical.

10 Steffy 1994: 49.

11 See Rogers 1996: 59.

Stage 6

The bracing timbers were attached to the hull to reinforce its weakest points according to the choice of the shipbuilder or its crew during use. In the central part of the hull, a pair of bracing timbers corresponding to the mast-step was joined to the proto-keel.

Stage 7

The ship was not decked, as shown by the irregular form and inclination of the beams, as well as by the complete absence of any surface traces of deck beams and planking. The centrally positioned mast and rigging were set up at this final stage of the construction (Figure 9.6).

The second book of Herodotus' *Historia* contains a very important fragment related to the topic:

> *Now in old times Naukratis alone was an open trading-place, and no other place in Egypt: and if any one came to any other of the Nile mouths, he was compelled to swear that he came not thither of his own will, and when he had thus sworn his innocence he had to sail with his ship to the Canopic mouth, or if it were not possible to sail by reason of contrary winds, then he had to carry his cargo round the head of the Delta in boats* ['baris' in the original text: AB] *to Naukratis: thus highly was Naukratis privileged.*[46]

It is significant that Herodotus uses different terms for foreign seagoing vessels (generic term of *ploion*) and for the ships employed for local transportation (*baris*). This indicates that the *barides* could operate beyond the Delta and thus belonged to a class of fluvio-maritime vessels.

The depth of the ports of Thonis-Heracleion is difficult to determine with precision, but in accordance with the morphology of coastal lagoons it hardly exceeded two-three metres on average.[47] Thus the shallow-drafted *barides* obviously were well-adapted to this environment. In addition, navigation on the Nile was highly seasonal[48] and their smaller specimens could operate for a longer time period than other types.[49]

An interesting clue is offered by the anchors that were found in great numbers (more than 700) in the harbours of Thonis-Heracleion. The anchors appear in different types, but most of them are triangular stone composite anchors with two round front openings for wooden flukes and one transverse opening for the cable.[50] The majority of anchors are about 70–90 cm long and approaching 100 kg. Some of these anchors were found on board of the *barides* in position relevant for mooring. This is definitely the case of ship 43 that had a 100 kg anchor placed in vertical position at the bow.[51] These anchors were probably handled with a help of tackles and a mast-derrick.[52]

While there is plenty of archaeological evidence for ancient Egyptian anchors from their sea-going vessels,[53] these were of no real use on the Nile.[54] Instead, a wooden stake[55] was driven into the muddy shore with a mallet,[56] as seen in iconographical sources.[57] There are also no solid iconographical proofs for the use of anchors by Nile ships. Nevertheless, it may be suggested that mooring techniques could change during the Late Period following the increase of maritime trade and fluvio-maritime traffic for which anchors were indispensable. Several finds of 'elongated composite stone anchors'[58] in Egypt were made in its riverine environment.[59]

The Delta was a specific region between the river and the sea characterised by its proper navigational conditions and hazards.[60] According to Yoyotte, the ancient name of the city—Thonis—(Θῶνις in ancient Greek sources) originates in the indigenous name of the coastal lagoon (*henet/hone*) that existed here in antiquity.[61] The water of the lagoon was only slightly brackish and this is confirmed by numerous finds of bones of Nile catfish (*Siluriformes*) and other fresh-water organisms.[62] The sedimentology of a coastal lagoon is very different to that of a river and includes sediments ranging from coarse sand to silt and clay.[63] Many hundreds of finds from Thonis-Heracleion prove that this environment allowed the regular employment of marine-type anchors. It is not surprising, taking into consideration the intense shipping that required complicated manoeuvres in the restricted harbour space. It appears that in the Ptolemaic Period the river was perceived as the extension of the sea[64] so that not

---

46 Herodotus *Historia* 2.179. 'ἦν δὲ τὸ παλαιὸν μούνη Ναύκρατις ἐμπόριον καὶ ἄλλο οὐδὲν Αἰγύπτου· εἰ δέ τις ἐς τῶν τι ἄλλο στομάτων τοῦ Νείλου ἀπίκοιτο, χρῆν ὀμόσαι μὴ μὲν ἑκόντα ἐλθεῖν, ἀπομόσαντα δὲ τῇ νηὶ αὐτῇ πλέειν ἐς τὸ Κανωβικόν ἢ εἰ μή γε οἷά τε εἴη πρὸς ἀνέμους ἀντίους πλέειν, τὰ φορτία ἔδεε περιάγειν ἐν βάρισι περὶ τὸ Δέλτα, μέχρι οὗ ἀπίκοιτο ἐς Ναύκρατιν. οὕτω μὲν δὴ Ναύκρατις ἐτετίμητο.' Herodotus does not mention Thonis. The hypothesis dealing with this omission can be found in Höckmann 2008–9: 115, 124.

47 Cataudella *et al.* 2015: 73, table 2.

48 Cooper 2011: 195; 2012a: 61; 2012b.

49 See Robinson 2015: 213. Cooper (2012b: 26) cites the nineteenth century sources according to which the ships with deadweight of 60 tonnes were not able to navigate in the Delta during five months of the year.

50 See Nibbi 1991.

51 Calibrated date ^{14}C for wooden flukes: 405 cal–208 cal BC. Dimensions: 75 × 50 × 18 cm. Cf. Robinson 2015: fig. 11.1.

52 Basch 1987: 66–7; Frost 1995. It cannot be excluded that the largest anchors from Thonis-Heracleion were used as mooring anchors. This idea was introduced by several members of jury for my PhD thesis on 31st of January 2014 (P. Arnaud, P. Pomey, F. Goddio). For example the largest of two anchors found at the bow of ship 51 weights 630 kg (calibrated date ^{14}C for wooden arms: 396 cal–198 cal BC. Dimensions: anchor 1,106 × 80 × 26 cm; anchor 2, 154 × 94 × 30 cm). The hull of this ship was preserved only partially but the dimensions of its planking indicate that the length of the ship did not exceed 20–25 m so it was not a ship of extraordinary size. The same destination was proposed for the pyramidal stone anchors from Zea (see Tzalas 1999). Bronze Age stone anchors weighing 850 and even 1,350 kg are known from Kition (see Frost 1985).

53 Basch 1985; 1994; Zazzaro 2007; 2011; Zazzaro and Abd el-Maguid 2012; Tallet 2013a; 2015.

54 Basch 1985: 457; 1994.

55 *mnit* or *nˁyt* (Jones 1988: 198 note 4, 199 note 8).

56 *ẖrpw* (Jones 1988: 201 note 12).

57 For ancient Egyptian iconography of mooring see Doyle 1998: 220–35.

58 Frost 1970: 381.

59 Abd el-Maguid 2015.

60 Cooper 2012a; 2012b.

61 *henet/hone* → T(hone) → Thonis. See Yoyotte 2001; 2013: 298–9, 307–8, 349–52. This specific geography is described in classical sources: Heliodorus *Aethiopica* 5; Achilles Tatius *Leucippe and Clitophon* 4.12.7–8; Diodorus Siculus *Bibliotheca Historica* 1.31.2–5. For discussion see Fabre 2015: 180–4.

62 Goddio 2007: 111.

63 For geomorphology of coastal lagoons see Bird 1994. According to El-Wakeel and Wahby (1970) 'the predominant type of sediment covering the bottom of the lake [Manzalah – note A.B.] is the complex type sand-silt-clay followed in abundance by the clayey sand and silty clay. There is a basinward increase in grain size of sediments.'

64 Arnaud 2015: 104–5.

only the structure of river administration, but mooring procedures in a port (*hormos*) followed the maritime model. These imposed the choice of the Mediterranean type of mooring[65] and the employment of the marine type of anchors that were also necessary for open mooring.[66] The fact that anchors of marine type were used in Thonis-Heracleion does not bring on its own conclusive evidence for the navigational environment of the *barides*. These anchors could have been used for mooring beyond the sandbar separating the estuary from the open sea, or only within the confines of the port.

Ship 17, being one of the largest specimens of the *barides* discovered in Thonis-Heracleion, would have a shallow draft of about 1.6 m,[67] which would have been a definite advantage for navigation on the river and within the shallow lagoons of the Delta.[68] Furthermore, the depth at the mouths of the Nile was not considerable.[69]

The navigational space of *baris*-type ships can not be determined without examining the question of the trans-shipment of goods in Thonis-Heracleion. In the Late Period (664–332 BC) the city, 'The Gates of the Sea',[70] controlled access to Egypt, supervised Greek ships in transit for Naukratis[71] and Memphis[72] and thus functioned as a customs station and *emporion*.[73] The port commanded the strategic point at the mouth of one of the most important branches of the ancient Nile and perfectly fitted the category of a 'transit point' between 'transport zones' suggested by Westerdahl.[74] The conditions of navigation radically change at 'transit points' that are usually associated with market places and suggest 'the reloading of cargo and the change of means of transport at a well-defined site [...] for an accompanying water or land transport in the new zone'.[75] Flat-bottomed vessels from Ostia known as *naves caudicaria*, and used for trans-shipment of goods from the sea ports up the Tiber, can serve as a parallel.[76]

Höckmann argued for the regular trans-shipment of cargo from seagoing vessels to river-faring *barides* at Thonis-Heracleion.[77] An opposing point of view was expressed by Villing who, relying on recent research,[78] suggested that the Canopic branch in Naukratis 'was wide and deep enough to accommodate Mediterranean seagoing ships all year round; trans-shipment at Thonis-Heracleion, as had sometimes been suspected, was thus not necessary'.[79] It is difficult to adhere entirely to this conclusion, however, even though written sources are corroborated by several maritime finds from Naukratis.[80] Thomas estimates the Canopic branch to be 5 m deep and *c.* 200 m wide near urban quarters and concludes that 'Canopic branch of the Nile was deep enough and navigable, likely all year round, for sea-going ships such as the Kyrenia'.[81] We may note, however, that the Kyrenia ship was a fourth-century BC Greek merchantman of a very modest size. This 25-tonner's estimated length was about 14.5 m and had a draft of only 1.40 m,[82] consequently it would not have encountered serious problems in reaching Naukratis if it could transverse the bar at the end of the Canopic branch. For some foreign vessels, Thonis-Heracleion would have been their ultimate destination, where their goods were taxed by officials from the King's House and then sold in the *emporion*. Whereas other vessels, once customs duties had been paid, may well have sailed for destinations further upriver.

The Persian Period Aḥiqar scroll contains a list of foreign ships that passed through the unnamed port of the Delta, most probably Thonis-Heracleion.[83] There are two major groups of ships on the list: the 'small' ships had a tonnage of about 40 tonnes and the 'large' ones—about 60 tonnes.[84] Here again one is in presence of relatively small merchantmen that could easily come upriver on their own.

65 Known today as 'med mooring' or 'Tahitian mooring', this technique implies setting a temporary anchor off the pier and then approaching it at a right angle. The vessel then runs two lines to a pier.
66 Arnaud 2015: 104.
67 See this volume, Chapter 11.
68 Interesting parallel is suggested by the Arab-tradition fishing boats of the lakes Manzalah (Gaubert and Henein 2015) and Borollos (Collet and Pomey 2015). These boats with shallow draft are perfectly adapted to traditional fishing inside a coastal lagoon. The dimensions of *lokkafa* of the lake Borollos described by Collet and Pomey has an overall length of 14.5 m, a beam of 5.5 m and a shallow draft of 0.6 m. The boat carries a lateen sail with a windage of 130 m^2. The average depth of the lake is 1–1.5 m and rarely reaches 2 m. The available documentation on a traditional type of the lake Mareotis (*mariotike*) does not contain any information on their construction, but these ships were probably shallow-drafted as well (*BGU* 18.1 2740, dated to 87–86 BC. See Arnaud 2015a: 111).
69 Cooper (2012a: 61) cites the late nineteenth century data according to which 'the Rashid mouth had a maximum draught of 2.1 m, and the Dumyât mouth just 1.8 m, compared with 6 m just upstream'.
70 For association of this epithet—'Les Portes de la Mer'—with Thonis see Yoyotte 1994: 683.
71 Yoyotte 1958: 427; 2001: 27.
72 Aramaic papyrus from Saqqara note 26. Yoyotte 1994: 683; Briant and Descat 1998: 93–5.
73 Fabre and Goddio 2013: 70. Thonis-Heracleion is the most probable site of taxation for the ships mentioned on the Papyrus palimpsest from Elephantine (Aḥiqar scroll, TAD C 3.7, 473–402 BC). See Briant and Descat 1998: 92. The papyrus contains customs register of thirty-six Ionian and Phoenician ships that came to Egypt in the period from March to December. See Porten and Yardeni 1993.
74 The conception of Westerdahl (1992: 6–7) in its application to the mouths of the Nile is discussed by Cooper (2012a: 70–1).
75 Westerdahl 1998.
76 Boetto 2001; 2003; 2006; 2008.
77 Höckmann 2008–9: 78–80, 82–3.
78 Pennington and Thomas, in preparation; Thomas 2015.
79 Villing 2015: 231.
80 Thomas 2015: 253, fig. 13.5.
81 Ibid.: 252.
82 For the tonnage see Steffy 198: 100, for the draft see Katzev 1981: 318.
83 Höckmann 2008–9: 110.
84 Briant and Descat 1998: 68.

Arnaud shows that in the Hellenistic Period large seagoing ships with tonnage exceeding 10,000 *artab* or about 300 metric tonnes were more numerous on the Nile in winter, from December to February, than during the flood.[85] These ships included a seagoing *kerkuros*[86]—probably the most significant type of the time that could reach a tonnage of 450 metric tonnes.[87] It is essential to remember that these ships operated from Alexandria connected by a canal to the Nile and so did not need to enter its mouths like in preceding periods.

As far as the *barides* of Thonis-Heracleion are concerned, their construction strongly indicates a river origin and probably a fluvio-maritime working environment.[88] If the assumption of the hull form of ship 17 is correct they were well-adapted to navigation within the estuary, but were not particularly seaworthy. Several possibilities can be proposed for their use: either they were involved in the trans-shipment of goods from the larger sea-going vessels that could not enter Thonis-Heracleion because of their considerable draft, or they transported goods from the city up the river, or, finally, both. Furthermore, they did not need to be involved in international maritime trade and could have belonged to a temple fleet that collected tithes from its land.[89]

It appears that the question of trans-shipment at Thonis-Heracleion is not as unambiguous as it is shown sometimes.[90] There were too many parameters involved: the tonnage of the ship, the nature of her cargo, seasonality and meteorological conditions, etc. Some of the seagoing ships continued their journey upriver on their own while others needed trans-shipment to the river-faring craft. It was simply a question of tonnage and draft and the conditions on the river. Herodotus makes it plain that it was sometimes difficult to navigate around the Delta due to the contrary winds'[91] and periods of calm winds could also lead to the necessity of trans-shipment.[92]

In conclusion it may be proposed that ship 17 belonged to a fleet of vessels of fluvio-maritime type described as *barides* by Herodotus that mainly operated on the Nile and in its Delta. Occasionally these vessels could go beyond the sand bar, most probably for the trans-shipment of goods.

85 Arnaud 2015: 106–9, 112.
86 See Thompson 1983: 73.
87 Arnaud 2015: 112.
88 Ibid.: 116.
89 Robinson 2015: 222, 291 note 51.
90 Cf. Höckmann 2008–9: 83 and Villing 2015: 231.
91 Herodotus *Historia* 2.179.
92 Cooper 2012b: 26. 'The warming land often generated land breezes that cancelled out the prevailing northwesterlies blowing in off the Mediterranean Sea, resulting in frequent calms that left Nile boats facing a strong current with no supporting wind. Even when the sea breezes broke through, Delta channels were not always oriented in a way that enabled navigators to take advantage of them. The combined result of all these factors was hard labour for Nile navigators'.

11 The main characteristics of the hull

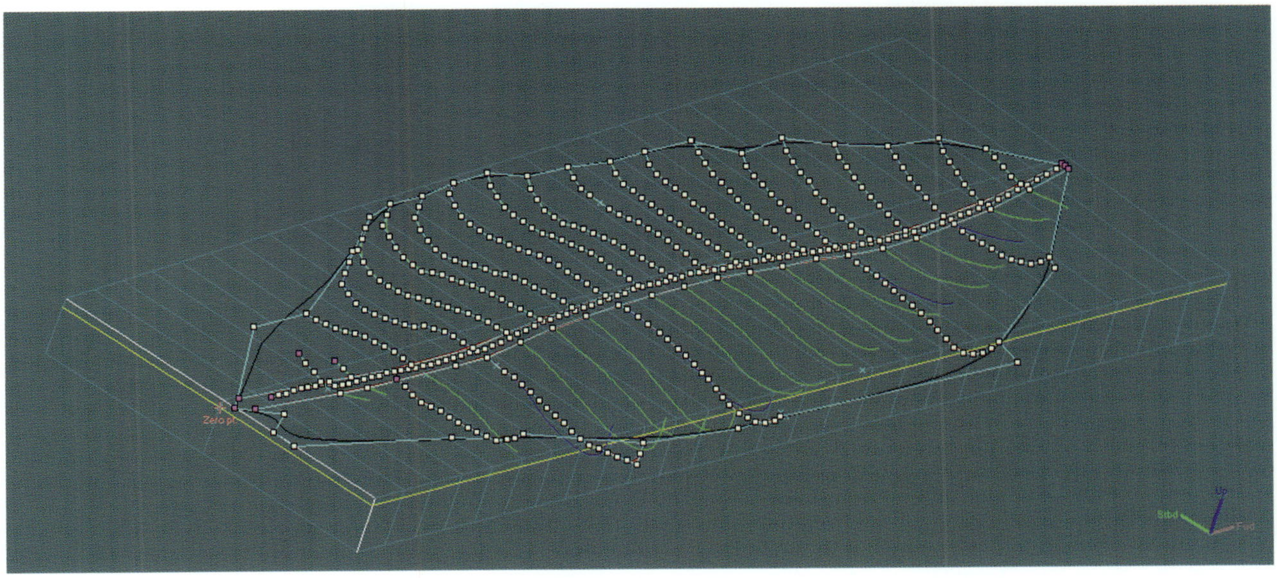

Figure 11.1 Position of the hull of ship 17 *in situ* according to data obtained with goniometer.

A preliminary version of the three-dimensional model for ship 17

Ship 17 provides valuable information about the construction details of a Nile freighter dating to the Late Period. Little is known of the particularities of ancient Egyptian shipbuilding during this period and thus reconstructing the shape of the hull of the ship and obtaining initial estimates for its technical characteristics is an attractive proposition.

Despite its good state of preservation, the reconstruction of ship 17 encounters serious problems. The sections of the proto-keel and the planks were characterised by their short length. At the same time the parts of the internal structure were neither sufficiently developed nor regular enough to serve as reliable indicators for reconstructing of the shape of the hull. Thus the hull may be compared to skin, flexible enough to take various forms after its deposition. Therefore, the first version of the model of the hull of ship 17 presents what may be regarded as a working hypothesis.[1]

As the corpus of iconographic material for the Late Period is poor, I was obliged to look for possible parallels among the representations dating back to the Third Intermediate Period (1069–664 BC) and even to the end of the eighteenth dynasty (1550–1295 BC). The most important conclusion that can be drawn on the basis of this material is that the vessels of these periods, river-faring and seagoing alike, were characterised *par excellence* by a crescent-shaped hull. These are vividly presented by Landström, who illustrates ships of the late New Kingdom that take into account the conventions of ancient Egyptian art.[2]

Of special interest is the recent identification of a *baris* among the boats appearing on the Nile mosaic from Palestrina, Italy *c.* 125 BC.[3] Despite the fact that this section of the mosaic was much restored, the drawing made before restoration allowed Pomey to identify the main features of ship's construction. The hull was crescent-shaped and relatively symmetrical with the stern being higher than the prow. The single mast was installed a little bit closer to the bow than to the centre of the vessel, and it was steered by an axial rudder. The ship seems to have been decked and had a large cabin built of removable panels.

During the documentation of the remains of ship 17, the three-dimensional position of the hull and of some elements of its internal structure (mainly through-beams) were taken with a goniometer (Figure 11.1).[4] *In*

1 The following software was employed for modelling: FormSys *MaxSurf*™ suite (*MaxSurf*, *WorkShop*, and *Hydromax*) and Autodesk *3ds Max*™.

2 Landström 1970: 134–9.

3 Pomey 2015b: 164–6.

4 Cozzi 1988.

Figure 11.2 Deep fracture in keel segment K6. (Photo: A. Belov; © F. Goddio/ IEASM.)

situ remains confirm the assumption of the crescent-shaped hull. It is worth noting the two symmetrical curves formed by the extremities of the ship, while the central part remains on the higher level (see Figure 2.14). Taking into consideration the homogeneous character of the surrounding sediment (dense grey clay), it seems logical to explain the form of the hull *in situ* by the gradual immersion of the heavy extremities after the decomposition of the ship's internal structure. A deep transversal fracture situated exactly in the middle of the hull was probably also caused by these processes (Figure 11.2). An absence of vertical posts and short segments of the keel assembled longitudinally with simple and rather 'flexible' tongue-and-groove joints corresponds well with the crescent shape of the hull witnessed by iconography.

It is notable that the large amount of knife-shaped planks used from the fifth strake onwards suggests that the boat width decreased fairly rapidly towards the extremities beyond about 6–7 and 21–22 m from the zero point of the hull. A shell of this kind fitted with a horizontal keel, however, could be closed only using vertical stem and stern posts—a feature apparently absent in the structure of ship 17.

The arguments in favour of a crescent form of the hull are numerous but it is difficult to determine its exact curvature. For this reason the template for the graphical reconstruction is that of Landström for the Nile freighter from the New Kingdom.[5]

The proto-keel and first five strakes of both boards were assembled with the first set of tenon-ribs. The rectangular section of these strakes and rectangular channels confirm that this part of the hull was flat. The flat bottom extended to strakes number 10/11 assembled by the second set of tenon-ribs, as suggested by the massive supports that still preserved the remains of vertical stanchions.

The turn of the bilge was probably situated near starboard strake S13. This region corresponds to the outboard extremities of supports and at the same time it is characterised by the elevated density of tenons that attached the bracing timbers to the planking (Figures 11.3–4). It may be suggested that additional tenons were necessary to keep bracing timbers in place in the region of the turn of the bilge. The tenon-ribs of the third set that assembled the strakes from 10–11 to 15–16 were probably bent to achieve the required shape. According to available data ancient Egyptian vessels with a hard chine were built exclusively in the Old Kingdom,[6] although there is no indication of a hard chine on ship 17.

An important parameter in estimating the width of the ship is the length of the through-beam B1 that was preserved completely. This beam, 7.57 m long (7.16 m in a straight line between its extremities), was found locked into its starboard mortise and close to its port mortise. The reconstruction of the section at 8–9 m from the zero point was based on the length of the beam and its position in the planking, on the form of the bracing timber F3P, and on the assumption of the flat bottom extending to strake 13 (Figures 11.5–11.6).

The length of the hull was roughly estimated by the convergence of the better-preserved parts of the bulwark, keeping in mind that the stern segment K1 although not

5 Landström 1970: 136–7.
6 These ships had an angular profile and bow and stern transoms.

See Landström 1970: 44–6, figs119–21; 66–8, figs 196–201.

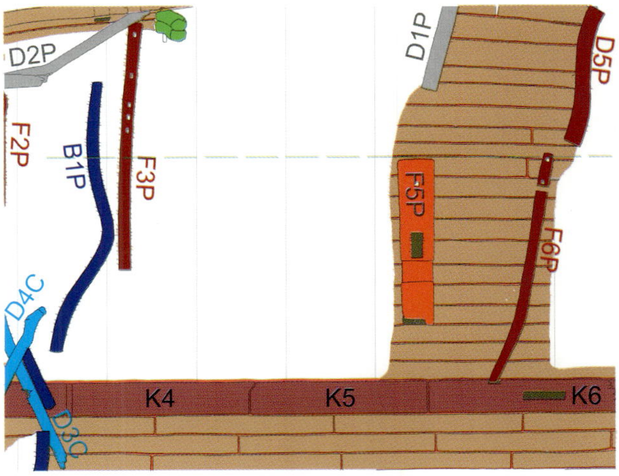

Figure 11.3 Central part of the hull on the port. Note that the beginning of the zone of high density of the tenons that attached bracing timbers to the planking corresponds to the outboard extremity of support F5P (green line). (Drawing: P. Sandrin/A. Belov; © F. Goddio/IEASM.)

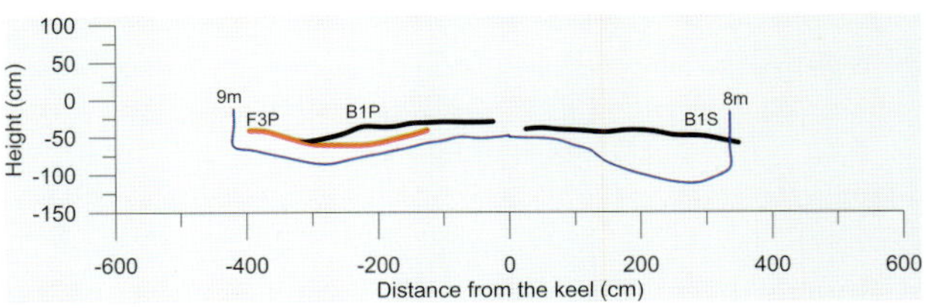

Figure 11.5 Available data on the section of the hull at 8–9 m from zero point. (Drawing: A. Belov.)

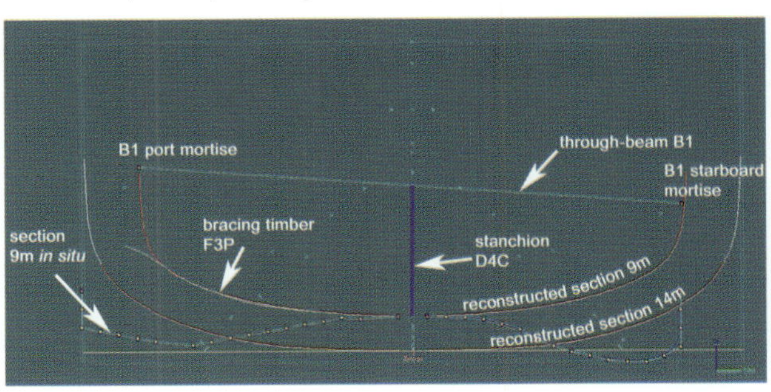

Figure 11.6 Reconstruction of the transversal form of the hull at 8–9 m and 14 m from zero point. (Drawing: A. Belov.)

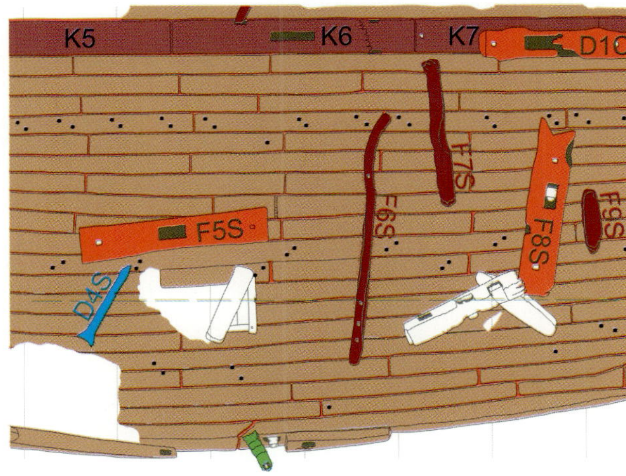

Figure 11.4 Central part of the hull on the starboard. Note that the beginning of the zone of high density of the tenons that attached bracing timbers to the planking corresponds to the outboard extremity of support F8S (green line). (Drawing: P. Sandrin/A. Belov; © F. Goddio/IEASM.)

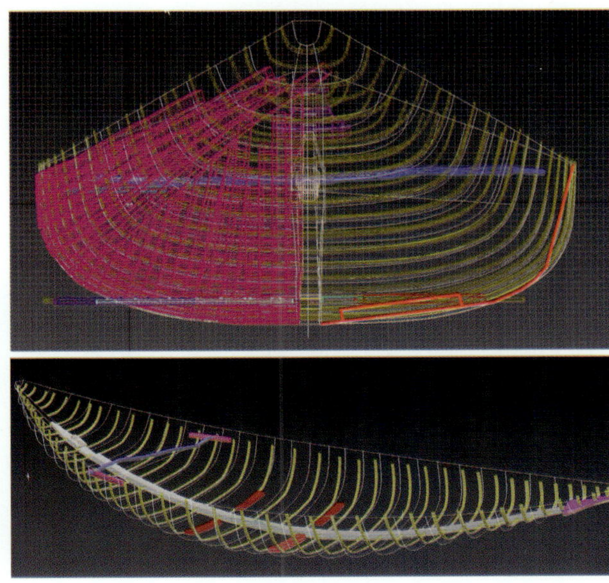

Figure 11.8 Model of the hull with a harder chine. (Graphic: Archéovision.)

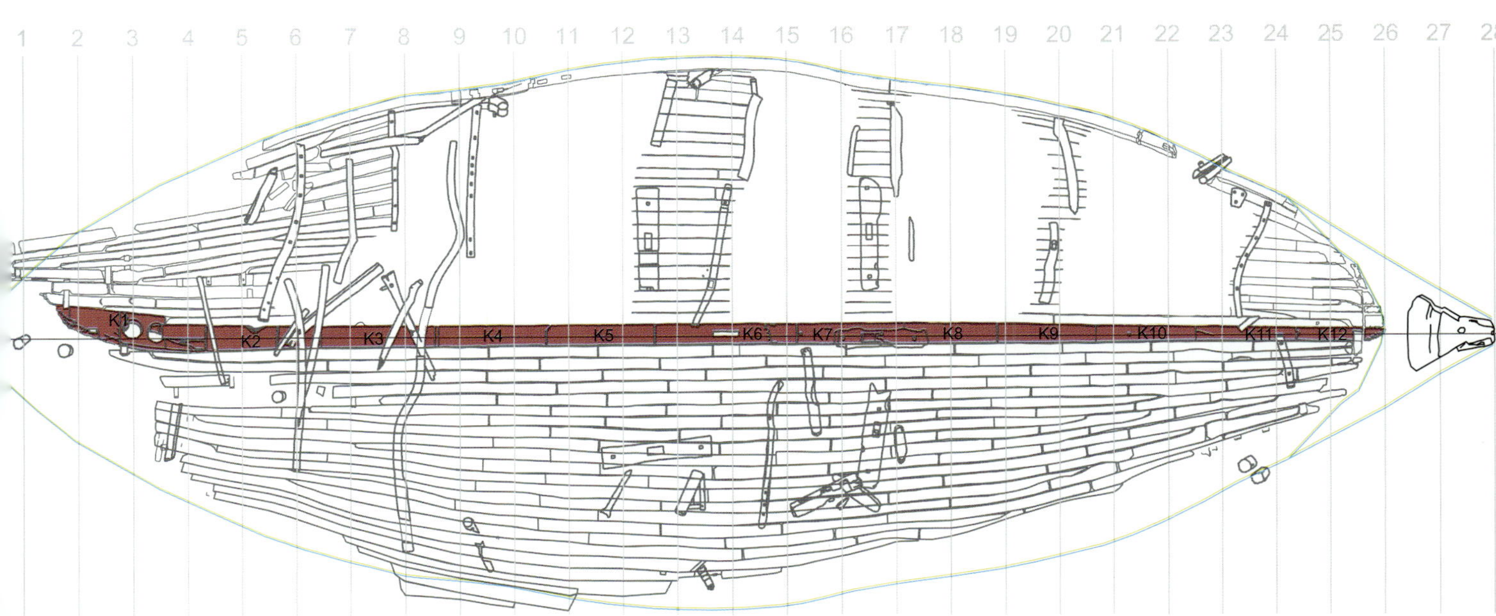

Figure 11.7 Evaluation of the length of the hull. (Drawing: P. Sandrin/A. Belov; © F. Goddio/IEASM.)

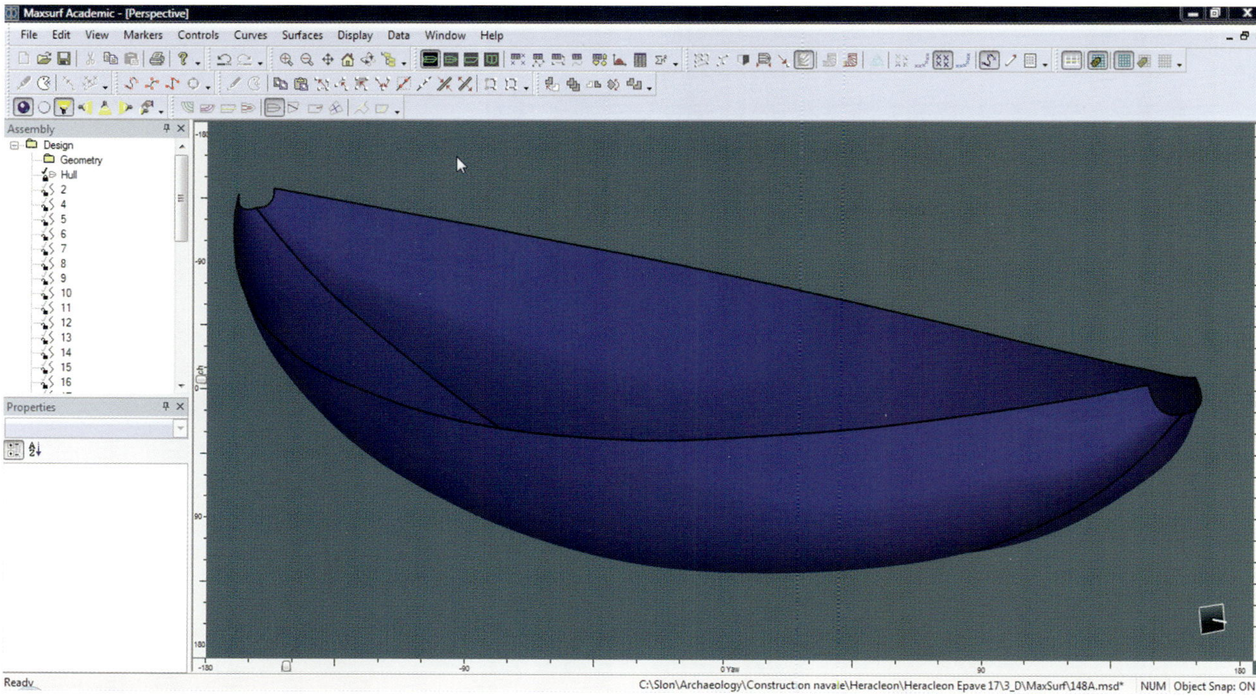

Figure 11.9 Axonometric view of the first version of the model. (Graphic: A. Belov.)

preserved completely was apparently the last one of the keel. The bow part of the ship was also badly preserved and was obviously lacking a 'bow piece' similar to one discovered on ship 43.[7] The dimensions and form of this detail were thus used in the reconstruction (Figure 11.7).

An alternative hypothesis with a hull shape with a harder chine was also considered.[8] It is, however, difficult to explain how the tenon-ribs of the planking were bent to achieve such a pronounced angle (Figure 11.8).

An average thickness of the planking of 14.1 cm was used, as was a density of 1170 kg/m^3 for the acacia wood from which the hull was made.

Form and dimensions of the hull

The results of the modeling suggest a sickle-shaped hull with considerable overhangs at the extremities of the vessel. The ship was characterised by a flat bottom, a flat keel, and a soft chine. The overall length of the ship would have been about 27–28 m with a beam of 8 m, which gives a length to width ratio of around 3:4 (Figure 11.9).[9]

Hydrostatics

Although the first version of the model of ship 17 is an initial approximation, it nevertheless allows an evaluation of the main characteristics of its hull. As the navigation area of ship 17 was probably limited to the Nile and the Delta, the density of fresh water (1.000 t/m^3) was used in the calculations.

The weight of the planking

The total surface of the hull of ship 17 was estimated at 205,408 m^2, with a volume of 28,963 m^3 [205.41 m^2 × 0.141 m (average thickness of the planking)]. The planking of acacia wood (density of 1170 kg/m^3) would thus weigh 33.89 metric tonnes.

The weight of details of internal structure and rigging

Supports

The average dimensions of the supports were 1.93 × 0.37 × 0.14 m that would give the volume of 0.099974 m^3 and a weight of 117 kg per piece. The total weight of the five supports would be 585 kg.

7 Ship 17 belongs to the same type of construction and has approximately the same dating as ship 17. See Robinson 2015.

8 This model was developed on *Autodesk '3ds Max'*™ software. I am very grateful to Robert Vergnieux and to the personel of *Archéovision* (UPS3551 SHS-3D) for their valuable help.

9 These figures fit fairly well into the data from written sources. The Nile freighters of the Old Kingdom seem to be more 'square'. Thus, *sḥt* and *s3ṯ* ships built of acacia wood by Weni (*c.* 2200 BC, the sixth dynasty) were 60 cubits long and 40 cubits wide (approximately 31 × 21 m, length to width ratio of 1.5). *The Story of the Shipwrecked Sailor* (probably the twentieth dynasty, 1985–1773 BC) states the 3.0 ratio for seagoing ships (120 by 40 cubits). See Grandet 1998: 35–7. At the beginning of the eighteenth dynasty Karnak obelisks of Thutmose I (1504–1492 BC) were transported on the barges 120 cubits long and 40 cubits wide (about 63 × 21 m, ratio of 3.c). See the biography of Ineni, Breasted 1906: vol.2, 43, §105. Analysis of the ostraca from Ramesseum (1279–1213 BC) mentioning the transportation of the blocks of sandstone allowed Kitchen to suggest a length to width ratio of 3.0 for these freighters (*c.* 1250 BC). See Kitchen 1991. It seems that this ratio becomes smaller for the grain carriers of the Ramesside Period (2.5), see Hornell 1943b. The dimensions of eight ships mentioned on the blocks from Karnak (45 by 15 cubits) give again the length to width ratio of 3.0, see Janseln-Winkeln 2007: 354–7.

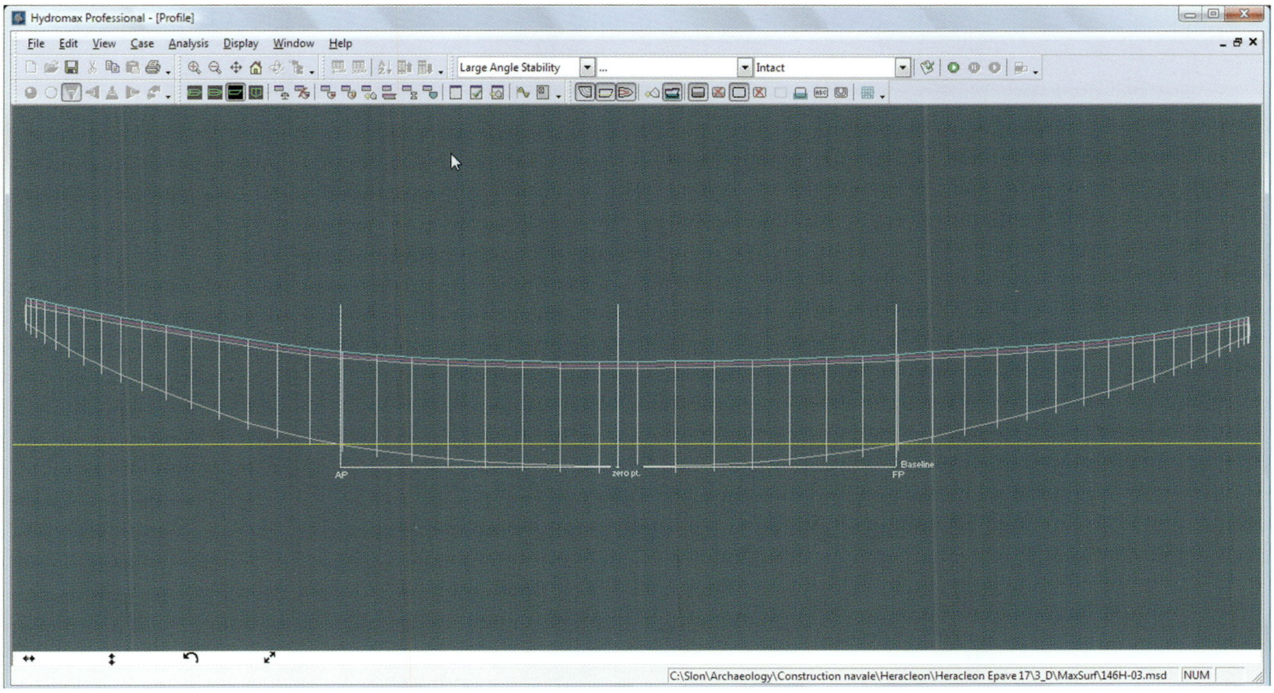

Figure 11.10 Sheer view of the model of the hull in Hydromax under lightship condition. (Graphic: A. Belov.)

The weight of the bracing timbers

The number of these pieces is estimated at twelve, whose average dimensions are 2.32 × 0.14 × 0.08 m. Thus the weight of one piece would be around 30 kg and the total weight around 360 kg.

The weight of the through-beams

The weight of the completely preserved beam B1, 7.57 m long and 8–15 cm wide, would be around 106 kg. The number of through-beams is estimated at seven with a total weight of 742 kg.

The weight of the mast and yard

The height of the mast can be estimated at 18 m. Taking an average diameter of 20 cm we obtain the weight of 1,323 kg for a mast of acacia wood. There should have been at least one yard with the weight close to that of a through-beam (estimated at 742 kg).

The weight of the rudder

The steering oar of the Khufu I ship, made of cedar, weighs 57 kg. It may be suggested that the rudder of acacia wood must have been heavier (about 100 kg).

The total weight of details of the internal structure and rigging can be thus estimated at 3.94 metric tonnes.

Light displacement

First of all, the light displacement of the ship was calculated as the sum of the weight of the planking (33.89 t), the details of internal structure and the rigging (3.95 t), which equals 37.85 metric tonnes. The importance of this is that it allows the draft of the ship to be determined without cargo, as well as its freeboard—the height of the hull over the ship's water line, which is a very significant parameter of the vessel's seaworthiness.

Lightship draft

The lightship draft was estimated at 0.62 m in fresh water (Figure 11.10).

Loaded displacement

The upright hydrostatics[10] experiments allowed a range of drafts to be tested to establish the greatest allowable one. Taking into consideration the significant size of the ship and the fact that it was undecked, the draft would correspond to about 1.6 m, with a freeboard of 0.6 m. Under these conditions corresponding to full load case the ship would have a displacement of about 150 metric tonnes (Figure 11.11).

Stability

The horizontal distance between the centre of gravity and a vertical line through the centre of buoyancy of a ship that is displaced from the upright position is called the righting arm.[11] The knowledge of this characteristic is important as it allows the righting moment—the force that tends to return the ship in its upright position—to be determined. Ship 17 was very stable, according to the

10 Carried out on *Hydromax*™ software.

11 For definition and graphical representation of hydrostatic characteristics that appear in this chapter see, for example, Biran and López-Pulido 2014 on the theory of hydrostatics.

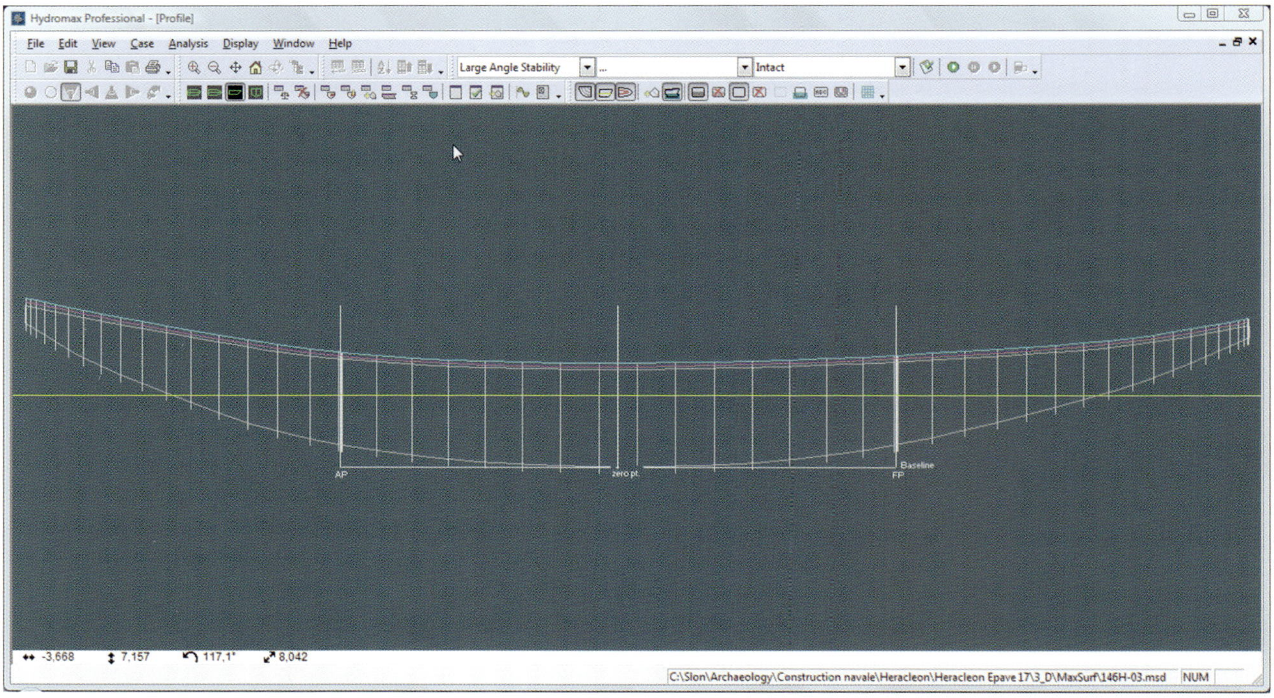

Figure 11.11 . Sheer view of the model of the hull in Hydromax when fully loaded. (Graphic: A. Belov.)

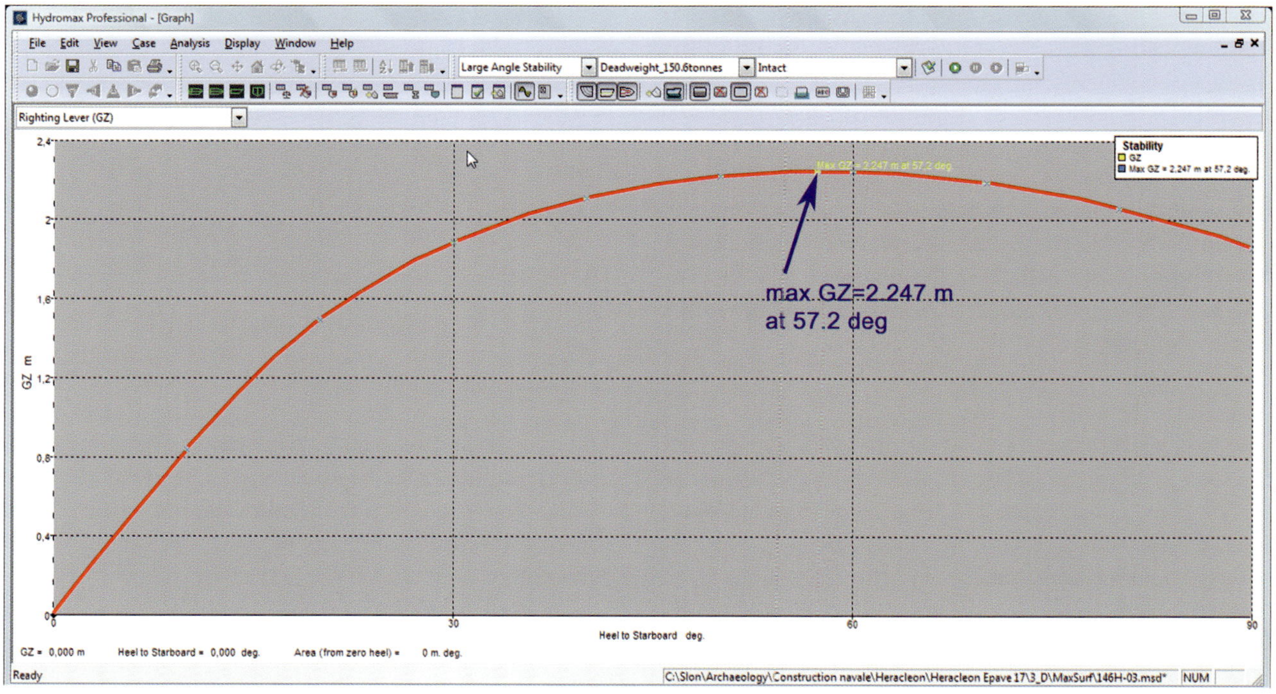

Figure 11.12 Curve of righting lever (GZ curve) when fully loaded. (Graphic: A. Belov)

results of this calculation, with the value of the righting arm (GZ) reaching its maximum at a heel[12] of 57.3 degrees. The absence of any deck, however, would limit the acceptable heel to only approximately 8 degrees (Figure 11.12).

Tonnage (deadweight)

The tonnage is calculated as the difference between the light displacement (37.9 metric tonnes) and loaded displacement (150.6 tonnes) and equals to 112.7 tonnes.

Main hydrostatic parameters

The upright hydrostatics characteristics are presented in Table 11.1. Due to the crescent-shaped hull of ship 17, its waterline length was only 66 per cent of its overall length. The hull was also full in section. The flat bottom of the ship explains the inconsiderable draft of 1.60 m, which was advantageous in the Nile and its estuary. This is probably why the proto-keel of the ship also did not protrude below the planking. The most ancient Egyptian

12 The sideways tilt of a ship under the influence of the wind and swell.

Table 11.1 Upright hydrostatics characteristics for the hull.

Characteristics	Abbreviation	Value	Unit
Overall length	LOA	27.16	m
Waterline length	LWL	18.02	m
Beam	B	8.04	m
Beam at waterline (WL)	LBP	7.73	m
Immersed depth, draft amidships (d or T)	T	1.60	m
Depth (D)	D	2.23	m
Max cross sectional area	A_{max} (Ax)	7.47	m^2
Wetted Surface Area	WSA	157.93	m^2
Water Plane Area	WPA	135.18	m^2
Displacement	Δ	150.6	tonne
Deadweight	DWT (tpl)	112.7	tonne

term we know of referring to a keel or to a keel plank ('pipit')[13] means a 'mud-kneader'.[14] The same type of flat keel is incorporated in the construction of the *nuggars* ethnographically recorded in the Upper Nile.[15]

The maximum cross sectional area A_x refers to the largest underwater section of a hull (midship section). For ship 17 this characteristic was estimated at 7.5 m^2, a significant figure reflecting the considerable size of the ship.

The water plane[16] area of the ship was also large at about 135 m^2. The rapid decrease of the hull's volume towards the extremities is reflected by the modest value of the block coefficient (0.54).[17] The hull was symmetrical and thus the longitudinal centre of flotation[18] and the longitudinal centre of buoyancy[19] are very close to the midship area. The transverse metacentre[20] (KMt) was situated at 4.5 m and indicates that the ship had good stability. The considerable displacement of 150 metric tonnes and a deadweight of 113 tonnes allow ship 17 to be classified as a large freighter (Table 11.2).

Table 11.2 Upright hydrostatics for ship 17 when fully loaded (112.7 metric tonnes).

Characteristic	Value
Draft Amidships, m	1.589
Displacement, tonne	150.6
Draft at Forward Perpendicular, m	1.583
Draft at After Perpendicular, m	1.595
Trim (positive by stern), m	0.012
Waterline Length, m	20.472
Waterline Beam, m	7.942
Wetted Area, m^2	157.995
Waterplane Area, m^2	135.368
Prismatic Coefficient	0.632
Block Coefficient	0.533
Midship Area Coefficient	0.843
Waterplane Area Coefficient	0.833
Longitudinal centre of buoyancy (LCB) from amidsh. (positive fwd), m	0.001
Longitudinal centre of flotation (LCF) from amidsh. (positive fwd), m	0.243
"Keel Buoyancy" (KB): vertical distance to vertical center of buoyancy (VCB), m	0.932
"Keel Gravity" (KG) is the distance from the baseline to the vessels's effective Vertical Center of Gravity (VCG), m	-0.349
Transverse metacentric height (GMt) corrected, m	4.837
Longitudinal metacentric height (GML) corrected, m	27.697
Transverse metacentre (KMt), m	4.488
Longitudinal metacentre values (KML), m	27.349
Immersion (TPc), tonne/cm	1.354
Moment to change trim one cm (MTc), tonne.m	2.314
Righting moment RM, tonne.m	12.711

Conclusions

The first version of the model of ship 17 presents a hypothesis for the probable form of its hull. It is likely that the ship had a sickle-shaped hull with considerable overhangs at the extremities of the vessel. Its bottom and keel were flat, which was advantageous for navigation within the estuary and on the Nile. The lightship draft of the ship was estimated at only 0.6 m in fresh water.

The overall length of the ship would have been about 27–28 m with a beam of 8 m that gives a length to width ratio of around 3.4. According to the upright hydrostatics experiments the ship in full load would have a draft of about 1.6 m, a freeboard of 0.6 m and displacement around 150 metric tonnes. The deadweight of ship 17 was near to 113 metric tonnes. These characteristics allow ship 17 to be classified among the largest freighters yet known from the site of Thonis-Heracleion.

Ship 17 was quite stable but the absence of deck would have limited the acceptable heel to approximately eight degrees.

It appears that ships of this type (*baris*)[21] were well-adapted for operating within a fluvio-maritime environment and could transport goods and passengers not only on the Nile and in its estuary but could also be involved in coastal trade.

13 Jones 1988: 164.
14 Goedicke 1975: 95; Janssen 1975: 379.
15 Clark 1920: 49; Hornell 1943a: 28.
16 The area at the intersection of the surface of the water and the waterline of a ship. It can vary according to the ship's draft.
17 The block coefficient is calculated as the volume of water displaced by a hull divided by the factum of its waterline length, waterline breadth and draft. Thus it represents the ratio between the volume of a box around the submerged part of the ship and actual volume occupied by the ship. Therefore, full forms will have a high block coefficient where fine shapes such as sailboats will have a low one.
18 The longitudinal centre of flotation (LCF) is the longitudinal distance from midships to the centre of the area of waterplane when the hull is not moving.
19 The longitudinal centre of buoyancy (LCB) is the longitudinal distance from midships to the centre of the displaced volume of water when the hull is not moving.
20 When a ship heels, the centre of buoyancy of the ship moves laterally. It might also move up or down with respect to the water line. The point at which a vertical line through the heeled centre of buoyancy crosses the line through the original, the vertical centre of buoyancy is the metacentre. The metacentre remains directly above the centre of buoyancy by definition.
21 See this volume, Chapter 10.

Conclusion

According to available radiocarbon and ceramic data, ship 17 most probably sank in the first half of the fifth century BC. Its construction shows numerous parallels with Herodotus' description of the local Egyptian river boat called the *baris*. The word-by-word analysis of this text demonstrates that all the statements of Herodotus, except for the smaller length of the planks and the absence of frames in its construction, correspond exactly to the evidence from ship 17. These two mismatches can be explained by the larger size of ship 17 in comparison with the boat described by Herodotus. The resemblance between the *baris* and ship 17 starts with the construction material (acacia) and the type of boat (Nile freighter) and then continues with both the longitudinal (keel, planking) and transversal (through-beams) structure of the ship. The methods of assembling the planking and making the hull watertight are exactly the same as in Herodotus' description. Finally, the steering system of ship 17 and the presence of the mast are also corroborated by the ancient text. The arguments cited above allow us to state that ship 17 from Thonis-Heracleion most probably belongs to a type of vessel that Herodotus described, or at least to a type that was closely related to the *baris*.

Ship 17 was constructed in a way that has not previously been seen archaeologically and analysis of it shows numerous features typical of ancient Egyptian shipbuilding: local timber used in construction, the crescent-shaped hull, the absence of conventional stem and stern posts, very thick planking and an axial rudder. On the other hand, ligatures, which remained a fairly typical feature of Egyptian naval architecture starting from the Old Kingdom, were completely absent in the construction of ship 17. The planks were assembled with very long tenons, a system that was quite unknown previously. While this form of building may find parallels in boatbuilding practices of the Old Kingdom, it could also be a later invention by ancient Egyptian shipwrights.

The form and position of the proto-keel of ship 17 finds parallels in the earlier Uluburun (*c.* 1300 BC)[1] and Gelidonia (*c.* 1200 BC) shipwrecks[2] belonging to the Syro-Canaanite shipbuilding tradition. The Uluburun ship is also the first one that uses pegged tenons for assembling the planks, which are equally found in construction of ship 17. The difference resides in the fact that the tenons of the Uluburun ship, although quite long, never exceed their own strake. It may be noted that the use of two-metre long tenons that were at the same time quite thick, was possible only within very thick planking, which was characteristic of ancient Egyptian ships made of acacia wood.

A first version of the three-dimensional model for ship 17 suggests a sickle-shaped hull with considerable overhangs at the extremities. The ship was characterised by a flat bottom, a flat keel, and a soft chine. The overall length of the ship would have been about 27–28 m with a beam of 8 m that gives a length to width ratio of around 3:4. The ship had a displacement of about 150 metric tonnes, a draft of 1.6 m and a tonnage of approximately 112 metric tonnes. This type of a boat with a shallow draft and a flat keel was perfectly suited to the navigational conditions of the Delta that included shallow waters and numerous sandbanks. Alternatively, these ships could have been used for the reloading of cargo onto large ocean-going freighters that were not capable of entering the port of Thonis-Heracleion because of their draft or of sailing upstream on their own.

Many aspects of the construction of a *baris* need additional research. The advantage of the site of Thonis-Heracleion is that the information provided by ship 17 can be supplemented and extended by the excavations of other ships dated to the Late Period. This will allow a more precise definition of the constructional principles and methods used during this time, many of which remain unique to the ancient Egyptian traditions of boatbuilding.

1 Pulak 2002: 618–21.
2 Bass 1967.

Glossary

Readers may also find it useful to consult:

Castro, F. (2017) *Library and Glossary of Early Mediterranean Shipwrecks for GROPLAN and iMARECULTURE. ShipLAB* Report 17 Version 05 – May 2017.

Falconer, W. (1780) *An universal dictionary of the Marine*, London.

Steffy, J. R. (2011) 'Illustrated Glossary of Ship and Boat Terms', *The Oxford Handbook of Maritime Archaeology*.

amidships – the middle of the ship, either with regard to her length or breadth.

athwartship – reaching across the ship, from one side to the other.

barge – a flat-bottomed vessel of burthen, for lading and discharging ships, and removing their cargoes from place to place in a harbour.

beam – strong thick piece of timber, stretching across the ship from side to side, to support the decks, and retain the sides at their proper distance.

bow – forward part of a hull; portion of the hull where the sides converge into the stem.

bulkhead – a partition or wall, built up in several places of a ship under the deck, either lengthwise or across, to form and separate the various apartments.

bulwark – portion of the side of a vessel above the upper deck.

caulk – to drive a quantity of oakum, or old ropes into the seams of the planks, or into the intervals where the planks are joined to each other in the ship's decks or sides, in order to prevent the ingress of water. After the oakum is driven very hard into these seams, it is covered with hot melted pitch or resin, to keep the water from rotting it.

chine – angular junction of the bottom and side of a vessel.

deadweight – deadweight tonnage (DWT) is a measure of a vessel's weight carrying capacity, and does not include the weight of the ship itself.

displacement – the weight of a ship, expressed in long tons of water (2,240 pounds) its hull displaces. It is measured indirectly using Archimedes' principle by first calculating the volume of water displaced by the ship then converting that value into tons.

draft – depth to which a hull is immersed.

floor timber – central and lower frame timber that spans the bottom of a vessel.

frame – transverse timber or assemblage of timbers fastened to the planking and sometimes to the keel that reinforce the hull.

freeboard – the distance between the waterline and upper deck.

futtock – frame timber that continues a floor timber or half frame upwards along the side of a hull.

garboard (strake) – in ship-building, the first range or strake of planks laid upon a ship's bottom next to the keel, throughout the whole length of the floor. The edge of this plank is let into a groove or channel in the side of the keel, which is called the rabbet of the garboard-strake.

grommet – a ring made of rope used for fixing the oars.

gunwale – the upper edge of a vessel's side.

half-frame – frame timber placed longitudinally in between floor timbers and covering the gap between floor timbers and futtocks.

half-lap joint (half-lap splice) – joint in which material is removed from both of the members so that the resulting joint is the thickness of the thickest member. Most commonly in half lap joints, the members are of the same thickness and half the thickness of each is removed.

hawse-holes – certain cylindrical holes cut through the bows of a ship on each side of the stem, through which the cables pass in order to be drawn into, or let out of the vessel, as occasion requires.

heel – to stoop or incline to either side. It is usually applied to a ship when she is forced into this position by the wind acting upon her sails, while braced obliquely across her, or by being ballasted to as to lean more to one side than the other.

hogging truss – in ancient vessels, a strong cable supported by forked posts and attached to the ends of the hull to prevent hogging.

hull – main structure of a ship's body, including the bottom and sides, but not the masts, spars, and rigging.

keel – longitudinal timber that forms the axis of a ship and supports the frames, deadwoods, and ends of the hull.

keel-plank – a central hull plank that was substantially thicker than the rest of the bottom planking and whose breadth was at least twice as great as its thickness; a thick bottom plank used in lieu of a keel.

lashing – when the planks are lashed together through holes, over a wadding of caulking material; unpegged tenons are inserted in mortises on the edges of the planks.
light displacement – (LDT) the weight of the ship excluding cargo, fuel, water, ballast, stores, passengers, crew.
lightship draft – the moulded draft amidships corresponding to the lightweight.
loaded displacement – the weight of the ship including cargo, passengers, fuel, water, stores, dunnage and such other items necessary for use on a voyage.
loose-footed sail – a sail without a boom.
luting – in most cases, animal hair, wool, or moss soaked in pitch or resin and laid in a luting cove, which was cut in the lower inside surface of the overlapping plank. Luting generally refers to caulking inserted between two hull members before they were assembled, as opposed to driven caulking.
mast-step – a mortise cut into the top of a keelson or large floor timber, or a mortised wooden block or assembly of blocks mounted on the floor.
mooring – the act of confining and securing a ship in a particular station, by chains or cables, which are either fastened to the adjacent shore, or to anchors on the bottom.
mortise-and-tenon joint – a form of edge fastening that consists of carving mortises on the top and bottom edges of the hull planks and regular intervals, inserting a wooden tenon in each lower and upper mortise, and locking them in place with two wooden pegs.
overhang – the part of a vessel's stern that projects aft of the rudder stock.
planking – outer lining of a hull.
port – the left-side of a ship when facing forward.
quarter-rudders – rudders mounted on one or both stern quarters; in use until the middle of the Medieval period.
rigging – a general name given to all the ropes employed to support the masts and to extend or reduce the sails, or to arrange them to the disposition of the wind.
rudder – a timber, or assembly of timbers, that could be rotated about an axis to control the direction of a vessel underway. Until the middle of the medieval period, the practice was to mount rudders on one or both stern quarters; these were known as quarter rudders.
rudder-stock – the part of a rudder which is closest to the sternpost. In relation to the axial rudder of ancient Egyptian ships it is the upper part of the steering oar that was supported by a stanchion.
scarf – overlapping joint connecting two timbers.
seaworthiness – sufficiency of a vessel for its intended voyage.
sheer view – the side view of a vessel's hull plan.
stanchion – a small pillar of wood or iron used to support decks, quarter-rails, nettings, awnings, etc.
starboard – the right side of a ship when facing forward.
stealer – a short plank inserted between two strakes of planking so that the regular strakes did not have to be made too wide; usually located at the bow or stern ends of bottom or lower side strakes.
stem – upward timber (or assembly) scarfed to the forward end of the keel (or central plank).
sternpost – upward timber (or assembly) scarfed to the aft end of the keel (or central plank).
strake – continuous line of planking in a hull. Planks forming strakes can be joined with diagonal scarves, S, or Z-shaped scarves, or simply butt joined.
superstructure – the parts of the ship or a boat that project above her main deck.
tackle – a machine formed by the communication of a rope with an assemblage of blocks, and known in mechanics as a pulley.
through-beams – an athwartships timber that extended through and beyond the outer hull planking. Through-beams were most common on ancient and Medieval hulls, where they supported the quarter rudders or provided athwartships stiffness to the upper part of the hull.
thwart – the seat or bench of a boat on which the rowers sit.
timbers – in general context, all wooden hull members; specifically, those members that formed the frames of a hull.
tongue-and-groove joint – a joint made between two boards by means of a tongue along the edge of one board that fits into a groove along the edge of the other board.
tonnage – a measure of the cargo-carrying capacity of a ship. Tonnage should not be confused with displacement, which refers to the actual weight of the vessel.
treenail – a long cylindrical wooden pin or dowel, employed to connect the planks of a ship's side and bottom to the corresponding timbers.
turn of the bilge – portion of the hull that divides the bottom and the sides.
wales – an assemblage of strong planks extending along a ship's side, throughout her whole length, at different heights, and serving to reinforce the decks, and forming the curves by which the vessel appears light and graceful on the water.
yard – a spar to which a square sail is set.

Bibliography

Classical sources

Aeschylus	*Supplices*
Euripides	*Iphigenia in Aulis*
Herodotus	*Historia*
Pliny the Elder	*Naturalis Historia*
Plutarch	*De Osiris et Isis*
Procopius	*De aedificiis*
Theophrastus	*Historia Plantarum*
Thucydides	*De Bello Peloponnesiacoa*

Abd El-Maguid, M. (2015) 'An elongated composite stone anchor from Matariya', in G. Soukkiassian (ed.), *Alexandria under the Mediterranean. Archaeological studies in memory of Honor Frost*. Alexandria: 125–37.

Arnaud, P. (2012) 'La mer, vecteur des mobilités grecques', in L. Capdetrey (ed.), *Mobilités grecques.* Bordeaux: 89–135.

Arnaud, P. (2015a) 'La batellerie de fret nilotique d'après la documentation papyrologique (300 av. J.-C. –400 apr. J.-C.)', in P. Pomey (ed.), *La batellerie Egyptienne. Archéologie, histoire, ethnographie*. Paris: 99–151.

Arnaud, P. (2015b) 'Navires et navigation commerciale sur la mer et sur le «Grand fleuve» à l'époque des Ptolémées', in B. Argémi and P. Tallet (eds), *Entre Nil et mers. La navigation en Égypte ancienne* (*NeHet* 2015 [3]): 105–22.

Assmann, E. (1913) 'Die schiffsbilder', in L. Borchardt (ed.), *Das Grabbenkmal des Königs Sahu-re*. Leipzig: 133–66.

Aston, D. A. (1999) *Pottery from the Late New Kingdom to the Early Ptolemaic Period*, Elephantine XIX. Mainz.

Barguet, P. (1986) *Les textes des sarcophages égyptiens du Moyen Empire. Introduction et traduction*. Paris.

Barguet, A., de Romilly, J. and Roussel, D. (1964) *Hérodote, Thucydide*. Paris.

Barker, R. (2014) 'Review of Kimura, J. (ed.) 2010, Shipwreck Asia: thematic studies in East Asian maritime archaeology. Adelaide', *International Journal of Nautical Archaeology* 43 (2): 479–81.

Barta, M. and Inglis, D. (2017) 'The Discovery of an Old Kingdom Boat at Abusir', Abstract of a paper presented at the *American Schools of Oriental Research* annual conference, Boston 2017. https://www.academia.edu/33227911/The_Discovery_of_an_Old_Kingdom_Boat_at_Abusir_ASOR_2017_Boston_Session_Maritime_Archaeology. Last consulted: 20 December 2018.

Basch, L. (1969) 'Phoenician oared ships', *Mariners Mirror* 55: 142.

Basch, L. (1972) 'Ancient wrecks and the archaeology of ships', *International Journal of Nautical Archaeology* 1: 1–58.

Basch, L. (1977) 'Triéres grecques, phéniciennes et égyptiennes', *Journal of Hellenic Studies* 97: 1–10.

Basch, L. (1978) 'Les navires *mnš et autres notes de voyage en Egypte*', *Mariners Mirror* 64: 115–8.

Basch, L. (1985) 'Anchors in Egypt', *Mariners Mirror* 71: 453–67.

Basch, L. (1986) 'Note sur le calfatage; la chose et le mot', *Archaeonautica* 6: 187–98.

Basch, L. (1987) *Le musée imaginaire de la marine antique*. Athens.

Basch, L. (1994) 'Some remarks on the use of stone anchors and pierced stones in Egypt', *International Journal of Nautical Archaeology* 23 (3): 219–27.

Basch, L. (1999) 'Le relief copte N9625 des Staatliche Museen de Berlin et le gouvernail d'étambot', in H. Tzalas (ed.), *TROPIS V—5th International Symposium on Ship Construction in Antiquity*. Athens: 43–51.

Basch, L. (2008) 'Recherche d'une généalogie', *Archaeologia Maritima Mediterranea* 5: 69–81.

Basch, L. (2015) 'Le calfatage: l'origine du mot et son aire d'extension', in P. Pomey (ed.), *La batellerie Egyptienne. Archéologie, histoire, ethnographie*. Paris: 227–35.

Bass, G. F. (1967) *Cape Gelidonya: A Bronze Age Shipwreck*. Philadelphia.

Bass, G. F. (1986). 'A Bronze Age Shipwreck at Ulu Burun (Kaş): 1984 Campaign', *American Journal of Archaeology* 90: 269–96.

Bell, C. D. J. (1933) 'Ancient Egyptian ship design, based on a critical analysis of the XIIth dynasty barge', *Ancient Egypt*: 101–11.

Bell, C. D. J. (1934) 'The obelisk barge of Hatshepsut', *Ancient Egypt*: 107.

Belov, A. (2014a) *Études de l'architecture navale égyptienne de la Basse Epoque. Nouvelle évidence archéologique et essai de restitution en 3D*. PhD thesis, Université Bordeaux Montaigne.

Belov, A. (2014b) 'A new type of construction evidenced by Ship 17 of Heracleion-Thonis', *International Journal of Nautical Archaeology* 43 (2): 314–29.

Belov, A. (2014c) 'New evidence for the steering system of the Egyptian *baris* (Herodotus 2.96)', *International Journal of Nautical Archaeology* 43 (1): 3–9.

Belov, A. (2015a) 'Did Ancient Egyptian Ships have Keels? The evidence of Thonis-Heracleion Ship 17', *International Journal of Nautical Archaeology* 44 (1): 74–80.

Belov, A. (2015b) 'Archaeological evidence for the Egyptian *baris* (Herodotus, II.96)', in D. Robinson and F. Goddio (eds), *Thonis-Heracleion in context*, Oxford Centre for Maritime Archaeology Monograph 8. Oxford: 189–204.

Belov, A. (2016) 'New light on the construction of the Egyptian *baris* as per Herodotus' narrative (2.96)', *Египет и сопредельные страны / Egypt and neighbouring countries* 1: 34–47.

Belov, A. (2018) 'Loose-footed sails of the Egyptian New Kingdom ships', *International Journal of Nautical Archaeology*. DOI: 10.1111/1095-9270.12335.

Belov, A. (forthcoming) 'A note on the navigation area of the *baris*-type ships of Thonis-Heracleion', in E. Nantet (ed.), *Proceedings of the international workshop Sailing from Polis to Empire: maritime scales in the Eastern Mediterranean during the Hellenistic period (4th–1st centuries* BCE*)*. Nicosia.

Benson, M. J. and Gourlay, J. (1899) *The temple of Mut in Asher. London.*

Bernand, A. (1970) *Le Delta égyptien d'après les textes grecs 1. Les confins libyques.* Le Caire.

Biran, A. and López-Pulido, R. (2014) *Ship Hydrostatics and Stability* (Second Edition). Oxford: 23–75.

Bird, E. C. F. (1994) 'Physical setting and geomorphology of coastal lagoons' in B. Kjerfve (ed.), *Coastal Lagoon Processes*. Amsterdam: 9–36.

Boetto, G. (2001) 'Les navires de Fiumicino', in J.-P. Descoeudres (ed.), *Ostia. Port et porte de la Rome antique*. Genève: 121–9.

Boetto, G. (2003) 'The Late-Roman *Fiumicino* 1 wreck: reconstructing the hull', in C. Beltrame (ed.), *Boats, ships and shipyards. Proceedings of the Ninth International Symposium on Boat and Ship Archaeology, Venice 2000*. Oxford: 66–70.

Boetto, G. (2006) *Les navires de Fiumicino (Italie): architecture, matérieaux, types et fonctions. Contribution à l'étude du système portuaire de Rome à l'époque impériale*. PhD thesis, Université de Provence.

Boetto, G. (2008) 'L'épave de l'Antiquité tardive Fiumicino 1: Analyse de la structure et étude fonctionnelle', *Archaeonautica* 15: 29–62.

Boreux, C. (1925) *Etudes de nautique égyptienne: l'art de la navigation en Egypte jusqu'à la fin de l'ancien Empire.* Le Caire.

Bradbury, L. (1996) '*Kpn*-boats, Punt Trade, and a Lost Emporium', *Journal of the American Research Center in Egypt* 33: 37–60.

Breasted, J. H. (1906–7). *Ancient records of Egypt.* Vols 1–5. Chicago.

Brewer, D. J. and Friedman, R. *Fish and fishing in Ancient Egypt.* Warminster.

Briant, P. and Descat, R. (1998). 'Un registre douanier de la satrapie d'Egypte à l'époque achéménide (TAD C 3. 7)', in N. Grimal (ed.), *Le commerce en Egypte ancienne.* Le Caire: 59–104.

Brugsch, H. K. (1868) *Hieroglyphisch-demotisches Wörterbuch. Leipzig.*

Bruni, S. (ed.) (2000) *Le navi antiche di Pisa: ad anno dall'inizio delle ricerche.* Florence.

Casson, L. (1942) 'Note on a Nile Boat', *The American Journal of Philology* 63 (3): 333–4.

Casson, L. (1950) 'The "Isis" and her voyage', *Transactions of the American Philological Association* 81: 43–56.

Casson, L. (1959) *The Ancient Mariners: Seafarers and Sea Fighters of the Mediterranean in Ancient Times.* New York.

Casson, L. (1990) 'Documentary evidence for Graeco-Roman shipbuilding (*P.Flor.* I 69)', *Bulletin of the American Society of Papyrologists* 27: 15–9.

Casson, L. (1995) *Ships and Seamanship in Ancient World.* Baltimore.

Casson, L. (2004) '"I've already sold my tunic": Nile skippers and their problems in the mid-third century B.C.', in F. Hocker (ed.), *The philosophy of shipbuilding. Conceptual approaches to the study of wooden ships.* College Station: 95–103.

Cataudella, S., Crosetti, D. and Massa, F. (eds) (2015) *Mediterranean coastal lagoons. Sustainable management and interactions among aquaculture, capture fisheries and the environment.* Rome.

Cervicek, P. (1974) *Felsbilder des Nord-Etbai, oberägyptens und unternubiens.* Wiesbaden.

Charlton, W. (1996) *Rope and the Art of Knot-Tying in the Seafaring of the Ancient Eastern Mediterranean.* PhD thesis, Texas A&M University.

Clarke, S. (1920) 'Nile boats and other matters', *Ancient Egypt*: 2–50.

Collet, R. and Pomey, P. (2015) 'Les voiles de Borollos', in P. Pomey (ed.), *La batellerie Egyptienne. Archéologie, histoire, ethnographie.* Paris: 299–315.

Cooper, J. P. (2008) *The Medieval Nile: Route, navigation and landscape in Islamic Egypt.* PhD thesis, University of Southampton.

Cooper, J. P. (2012a) '"Fear God; Fear the Bogaze": The Nile Mouths and the Navigational Landscape of the Medieval Nile Delta, Egypt', *Al-Masaq: Islam and the Medieval Mediterranean* 24 (1): 53–73.

Cooper, J. P. (2012b) 'Nile Navigation: "towing all day, punting for hours"', *Egyptian Archaeology* 41: 25–7.

Cooper, J. P. (2014) *The Medieval Nile: Route, Navigation, and Landscape in Islamic Egypt.* Cairo.

Cowley, A. (1923) *Aramaic papyri of the Vth century* BC. Oxford.

Cozzi, J. (1988) 'The Goniometer: an improved device for recording submerged shipwreck timbers', *International Journal of Nautical Archaeology* 27 (1): 64–80.

Creasman, P. P. (2010) 'A further investigation of the Cairo Dahshur boats', *Journal of Egyptian Archaeology* 96: 101–23.

Creasman, P. P. (2013) 'Ship timber and the reuse of wood in Ancient Egypt', *Journal of Egyptian Archaeology* 6: 152–76.

Darnell, J. C. (1992) 'The *Kbn.wt* vessels of the Late Period', in J. H. Johnson (ed.), *Life in a Multicultural Society: Egypt from Cambyses to Constantine and beyond*. Chicago: 67–89.

Davies, N. De Garis (1901–1902) *The rock tombs of Deir el-Gebrawi*. London.

Davies, N. De Garis (1920) *The tomb of Antefoker*. London.

Davies, N. De Garis (1926) *The tomb of Huy*. London.

Davies, N. De Garis (1927) *Two Ramesside tombs at Thebes*. New York.

Davies, N. De Garis (1935) *Paintings from the tomb of Rekh-mi-re' at Thebes. New York.*

Defernez, C. (2003) *La céramique d'époque perse à Tell El-Herr. Etude chrono-typologique et comparative*, Cahiers de recherches de l'Institut de papyrologie et égyptologie de Lille Suppl. 5. Lille.

Defernez, C. (2011) 'Les témoignages d'une continuité de la culture matérielle saïte à l'époque perse: l'apport de l'industrie céramique', in D. Devauchelle (ed.), *La XXVIe dynastie: continuités et ruptures. Actes du Colloque international organisé les 26 et 27 novembre 2004 à l'Université Charles-de-Gaulle-Lille 3. Promenade saïte avec Jean Yoyotte*. Lille: 109-26.

D'Oriano, R. and E. Riccardi, E. (2004) 'Les épaves d'Olbie', in S. Kingsley (ed.), *Barbares en Méditerranée de la Rome tardive au début de l'Islam*. Londres: 89–95.

Doyle, N. (1998) *Iconography and interpretation of Ancient Egyptian watercraft*. PhD thesis, Texas A&M University.

Edgerton, W. (1922–1923) 'Ancient Egyptian Ships and Shipping', *Asian Journal of Shipping and Logistics* 39: 109–29.

El-Wakeel, S. K. and Wahby, S. D. (1970) 'Bottom Sediments of Lake Manzalah, Egypt: notes', *Journal of Sedimentary Petrology* 40 (1): 480–96.

Emanuel, J. (2012) 'Egypt, the "Sea Peoples", and the Brailed Sail: Technological Transference in the Early Ramesside Period?', paper presented at *American Schools of Oriental Research Annual Meeting*, Chicago, IL.

Erichsen, W. (1954) *Demotisches Glossar*. Kopenhagen.

Fabre, D. (2005) *Seafaring in Ancient Egypt*. London.

Fabre, D. (2008) 'Heracleion-Thonis: Customs Station and Emporion', *Egypt's Sunken Treasures*. Munich: 219–36.

Fabre, D. (2011) 'The shipwrecks of Heracleion-Thonis: a preliminary study', in D. Robinson and A. Wilson (eds), *Maritime archaeology and ancient trade in the Mediterranean*, Oxford Centre for Maritime Archaeology Monograph 6. Oxford: 13–33.

Fabre, D. (2015) 'The ships of Thonis-Heracleion in context', in D. Robinson and F. Goddio (eds), *Thonis-Heracleion in context*, Oxford Centre for Maritime Archaeology Monograph 8. Oxford: 175–94.

Fabre, D. and Belov, A. (2012) 'The Shipwrecks of Heracleion-Thonis: An Overview', in G. A. Belova (ed.), *Achievements and problems of modern Egyptology. Proceedings of the international conference. September 29–October 4, 2009, Moscow*. Moscow: 107–18.

Fabre, D. and Goddio, F. (2013) 'Thonis-Heracleion, *emporion* of Egypt, recent discoveries and research perspectives: the shipwrecks', *Journal of Ancient Egyptian interconnections* 5 (1): 68–75.

Fagg, C. W. and Mugedo, J. Z. A. (2005) 'Acacia nilotica (L.) Willd. ex Delile', in P. C. M. Jansen (ed.) *PROTA 3: Dyes and tannins/Colorants et tanins. [CD-Rom]*. Wageningen.

Faulkner, R. O. (1940) 'Egyptian Seagoing Ships', *The Journal of Egyptian Archaeology* 26: 3–9.

Février, J. G. (1935) 'Les origines de la marine phénicienne', *Revue d'histoire de la philosophie et d'histoire général de la civilisation* 3: 97–124.

Foerster, F. (1989) 'Caulking in sewn boats?', *International Journal of Nautical Archaeology* 18 (3): 266–7.

Ford, B. (2013) 'The reuse of vessels as harbor structures: a cross-cultural comparison', *Journal of Maritime Archaeology* 8 (2): 197–219.

Frost, H. (1970) 'Bronze age stone anchors from the Eastern Mediterranean', *The Mariner's Mirror* 56: 377–94.

Frost, H. (1985) 'The Kition Anchors, in V. Karageorghis and M. Demas, *Kition V: The Pre-Phoenician Levels.* Nicosia: 281–321.

Frost, H. (1995) 'Where did bronze age ships keep their stone anchors?', in H. Tzalas (ed.), *TROPIS III. 3nd International Symposium on Ship Construction in Antiquity. Proceedings*. Athens: 167–76.

Gaubert, C. and Henein, N. H. (2015) 'Le bateau du lac Manzala', in P. Pomey (ed.), *La batellerie Egyptienne. Archéologie, histoire, ethnographie*. Paris: 285–99.

Glanville, S. R. K. (1930) 'Records of a Royal Dockyard of the time of Tuthmosis III: Papyrus British Museum 10056. Part I', *Zeitschrift für Ägyptische Sprache* 66: 105–21.

Glanville, S. R. K. (1932) 'Records of a Royal Dockyard of the time of Tuthmosis III: Papyrus British Museum 10056. Part II', *Zeitschrift für Ägyptische Sprache* 68: 7–41.

Goddio, F. (2007) *The Topography and the excavation of Heracleion-Thonis and East Canopus*

(1996–2006), Oxford Centre for Maritime Archaeology Monograph 1. Oxford.

Goddio, F. (2011) 'Heracleion-Thonis and Alexandria, two ancient Egyptian emporia', in D. Robinson and A. Wilson (eds), *Maritime Archaeology and Ancient Trade in the Mediterranean*, Oxford Centre for Maritime Archaeology Monograph 6. Oxford: 121–37.

Godley, A. D. (1921) *Herodotus' History*. London.

Goedicke, H. (1975) *The Report of Wenamun. Baltimore.*

Goyon, G. (1971) 'Les navires de transport de la chaussée monumentale d'Ounas', *Bulletin de l'institut français d'archéologie orientale* 69: 11–41.

Grandet, P. (1998) *Contes de l'Égypte ancienne.* Paris.

Graser, B. (1869) 'Das Seewesen der alten Aegypter', *Journal of Near Eastern Studies* 16: 4–13.

Grataloup, C. (2009) *Héracleion et Alexandrie. Rapport de l'étude céramologique.* Unpublished report. Alexandrie.

Grataloup, C. (2010) *Héracleion et Alexandrie. Rapport de l'étude céramologique.* Unpublished report. Alexandrie.

Grataloup, C. (2014) 'Les conclusions de l'étude céramologique', in A. Belov, *Études de l'architecture navale égyptienne de la Basse Epoque. Nouvelle évidence archéologique et essai de restitution en 3D.* PhD thesis, Université Bordeaux Montaigne: 418–21.

Guibal, F. and Pomey, P. (2002) 'Essences et qualité des bois utilisés dans la construction navale antique. L'apport de l'étude anatomique et dendrochronologique', *Forêt Méditerranéenne* 23 (2): 91–104.

Gunsenin, N. (2010) 'Introduction to the Shipwrecks at Yenikapı (İstanbul). The Byzantine harbour through history', in P. Pomey (ed.), *Transferts technologiques en architecture navale méditerranéenne de l'Antiquité aux temps modernes: identité technique et identité culturelle. Actes de la Table Ronde Internationale d'Istanbul, 19–21 mai 2007.* Istanbul: 107–08.

Haldane, C. W. and Shelmerdine, C. W. (1990) 'Herodotus 2.96.1–2 again', *The Classical Quarterly* 40 (2): 535–9.

Haldane, C. W. (1992) 'The Lisht timbers: A report on their significance', in D. Arnold (ed.), *The pyramid complex of Senwosret* I. New York: 102–12.

Haldane, C. W. (1993) *Ancient Egyptian hull construction*. PhD thesis, Texas A&M.

Haldane, C. W. (1996a) 'Ancient Egyptian hull construction', in H. Tzalas (ed.), *TROPIS IV. Proceedings of the 4th International Symposium on Ship Construction in Antiquity*. Athens: 235–44.

Haldane, C. W. (1996b) 'Mataria boat', in E. M. Meyers (ed.), *Encyclopaedia of the Ancient Near East.* Oxford: 441.

Hasslöf, O. (1963) 'Wrecks, archives, and living tradition: topical problems in marine-historical research', *Mariner's Mirror* 49: 163–4.

Hauben, H. (1979) 'Le transport fluvial en Egypte pharaonique. Les bateaux du roi et de la reine', *Actes du XVe Congrès International de Papyrologie*. Brussels: 68–77.

Helck, W. (1962) *Die Beziehungen Agyptens zu Vorderasien im 3. und 2. jahrtausend v. Chr.* Wiesbaden.

Hocker, F. (1998) 'Did Hatshepsut's Punt ships have keels?', in S. Wachsmann (ed.), *Seagoing Ships and Seamanship in the Bronze Age Levant*. London: 245–6.

Höckmann, O. (2008–2009) 'Griechischer Seeverkehr mit dem archaischen Naukratis in Ägypten', *Talanta* 40–41: 73–135.

Hornell, J. (1942) 'Frameless boats of the Middle Nile', *Sudan Notes and Records* 25: 1–36.

Hornell, J. (1943a) 'The sailing ship in Ancient Egypt', *Antiquity* 17: 27–41.

Hornell, J. (1943b) 'On the carrying capacity of Ramesside grain-ships', *Journal of Egyptian Archaeology* 29: 76–8.

Hudson, N. (2016) 'Late Persian and early Hellenistic pottery at Tell Timai', *Bulletin de liaison du groupe international d'étude de la céramiqiue* égyptienne 26: 75–108.

IFAO (2012). *Rapport d'activité 2011–2012.* http://www.ifao.egnet.net/uploads/rapports/Rapport_IFAO_2011-2012.pdf. Last consulted 30.04.2018.

Ingram, R. M. and Jones, M. R. (2010) 'Yenikapı: Documenting Two Byzantine Merchant Ships from the Yenikapı Excavations in Istanbul, Turkey', *INA Annual*: 8–17.

Janseln-Winkeln, K. (2007) *Inschriften der Spätzeit. Die 22.–24. Dynastie.* Wiesbaden.

Janssen, J. (1975) *Commodity prices from the Ramesside period.* Leiden.

Jéquier, G. (1911) 'Essai sur la nomenclature des parties des bateaux', *Bulletin de l'Institut français d'archéologie orientale* 9: 37–82.

Jones, D. (1988) *A Glossary of Ancient Egyptian Nautical Titles and Terms.* London.

Jones, D. (1990) *Model Boats from the Tomb of Tutankhamun.* Oxford.

Jones, D. (1995) *Egyptian bookshelf: Boats.* London.

Junker, H. (1934–1941). *Giza.* Vienna.

Kahanov, Y. (2011) 'The Dor/Tantura shipwrecks: clues to their construction tradition', *Archaeologia Maritima Mediterranea* 8: 137–51.

Kahanov, Y. and Mor, H. (2006) 'The Dor 2001/1 wreck. Dor/Tantura lagoon, Israel: preliminary report', in L. Blue (ed.), *Connected by the Sea. Proceedings of the Tenth International Symposium on Boat and Ship Archaeology Roskilde, 2003.* Oxford: 84–8.

Kahanov, Y. and Mor, H. (2009) 'Dor 2001/1: updated information and the retrieval of a section of the shipwreck', in R. Bockias (ed.), *Between the Seas. Transfer and exchange in Nautical Technology. Proceedings of the Eleventh International Symposium on Boat and Ship Archaeology, Mainz, 2006, ISBSA* 11. Mainz: 17–24.

KATZEV, M. L. (1981) 'The Reconstruction of the Kyrenia Ship, 1972–1975', *National Geographic Society Research Reports* 13: 315–28.

KITCHEN, K. A. (1991) 'Building the Ramesseum', *Cahiers de recherche de l'Institut de papyrologie et égyptologie de Lille* 13: 85–93.

KOCABAŞ, I. O. and KOCABAŞ, U. (2008) 'Technological and constructional features of Yenikapı shipwrecks: a preliminary evaluation', in U. Kocabaş (ed.), *The "Old ships" of the "New Gate"*. Istanbul: 97–186.

KOCABAŞ, U. (2014) 'Le port de Théodose et les 37 épaves de Yenikapı à Istanbul', *Dossiers d'Archéologie* 364: 78–86.

KOEHLER C. G. (1992) 'A Brief Typology and Chronology of Corinthian Transport Amphoras', in V. I. Kac and S. I. Monakhov (eds), *Les amphores grecques. Les problèmes de l'évolution du métier et du commerce dans l›antiquité.* Sarakoy: 265–83.

KOEHLER C. G. (1982) 'Amphoras on Amphoras', *Hesperia* 51: 284–92.

KÖSTER, A. (1923) 'Zur Seefahrt der alten Ägypten', *Zeitschrift für ägyptische Sprache und Altertumskunde* 58: 125–32.

LANDSTRÖM, B. (1969) *Sailing Ships in Words and Pictures from Papyrus Boats to Full-Riggers.* London.

LANDSTRÖM, B. (1970) *Ships of Pharaohs: 4000 Years of Egyptian Shipbuilding.* London.

LARCHER, P. and PESSONNEAUX, E. (1889) *Hérodote. Histoires.* Paris.

LEHMANN G. (1996) *Untersuchungen zur späten Eisenzeit in Syriaen und Libanon, Stratigraphie und Keramikformen swischen ca 720 bis 300 v Chr.* Münster.

LEWIS, N. (1960) 'On timber and Nile shipping', *Transactions of the American Philological Association* 91: 137–41.

LIDDELL, H. G. and SCOTT, R. (1940) *A Greek-English Lexicon revised and augmented throughout by Sir Henry Stuart Jones with the assistance of Roderick McKenzie.* Oxford.

LIPKE, P. (1984) *The Royal Ship of Cheops.* Oxford.

LLOYD, A. B. (1972a) 'Triremes and the Saite Navy', *The Journal of Egyptian Archaeology* 58: 268–79.

LLOYD, A. B. (1972b) 'The so-called galleys of Necho', *The Journal of Egyptian Archaeology* 58: 307–8.

LLOYD, A. B. (1975a) 'Were Necho's Triremes Phoenician?', *Journal of Hellenic Studies* 95: 55.

LLOYD, A. B. (1975b) *Herodotus, Book II. Introduction.* Leiden.

LLOYD, A. B. (1976) *Herodotus, Book II. Commentary 1–98.* Leiden.

LLOYD, A. B. (1979) 'Herodotus 2.96.1–1', *The Classical Quarterly* 29: 48.

LLOYD, A. B. (1980) 'The Trireme controversy. M. Basch on triremes: some observations', *Journal of Hellenic Studies* 100: 195–8.

MACAULAY, G. C. (1890) *The History of Herodotus.* London.

MAHFOUZ, E. (2012) 'New epigraphic material from Wadi Gawasis', in P. Tallet (ed.), *The Red Sea in pharaonic times: Recent discoveries along the Red Sea coast. Proceedings of the colloquium held in Cairo/Ayn Sukhna 11–12 January 2009*. Cairo.

MARK, S. (2009) 'The construction of the Khufu I vessel (*c.* 2566 BC): a re-evaluation', *International Journal of Nautical Archaeology* 38 (1): 133–52.

MARK, S. (2011) 'New data on Egyptian construction methods in the Khufu I vessel (*c.* 2566 BC) from the Paul Lipke collection', *International Journal of Nautical Archaeology* 40 (1): 18–38.

MARK, S. (2012) 'The Abydos BG10 boat and implications for standardization, innovation, and timber conservation in Early Dynastic boat-building', *Journal of Egyptian Archaeology* 98: 107–26.

MARK, S. (2017) 'The Ship Depiction in the Tomb of Nebamun: The Earliest Egyptian Ship Without a Hogging Truss', *Journal of Ancient Egyptian Interconnections* 16: 68–86.

MARX, E. (1947) 'Egyptian shipping', *Mariner's Mirror* 33: 139–58.

MASSON A. (2011) 'Persian and Ptolemaic Ceramics from Karnak: change and continuity', *Cahiers de la céramique égyptienne* 9: 39–310.

MCGRAIL, S. (2008) 'Sea Transport. Part I: Ships and Navigation', in J. P. Oleson (ed.), *The Oxford handbook of engineering and technology in the classical world.* New York: 606–37.

MCGRAIL, S. (2013) 'The transition from shell to skeleton in the Mediterranean and in North-west European waters', *International Journal of Nautical Archaeology* 42 (1): 188–9.

MEEKS, D. (1994) 'Compte rendu de D. Jones, A Glossary of Ancient Egyptian Nautical Titles and Terms (Londres, 1988)', *Chronique d'Egypte* 59 fasc.138: 254–60.

MERRIMAN, A. (2011) *Egyptian watercraft models from the Predynastic to Third Intermediate Periods.* Oxford.

MIDANT-REYNES, B. (2000) 'The Naqada Period (*c.* 4000–3200 BC)', in I. M. E. Shaw (ed.), *The Oxford History of Ancient Egypt.* Oxford.

MONTET, P. (1925) *Les scènes de la vie privée dans les tombeaux égyptiens de l'Ancien Empire. Strabourg.*

MOR, H. (2010) 'The Dor 2001/1 shipwreck: hull construction report', in P. Pomey (ed.), *Transferts technologiques en architecture navale méditerranéenne de l'Antiquité aux temps modernes: identité technique et identité culturelle. Actes de la Table Ronde Internationale d'Istanbul, 19–21 mai 2007.* Istanbul: 86–95.

MORRISON, J. S. (1972) 'Review of Casson L., *Ships and seamanship in the Ancient World*', *International Journal of Nautical Archaeology* 1: 230.

MORRISON, J. S. (1976) 'The classical tradition', in B. Greenhill (ed.), *Archaeology of the ship.* London: 155–73.

Muckelroy, K. (1976) 'The integration of historical and archaeological data concerning an historic wreck site: The "Kennemerland" ', *World Archaeology* 7 (3): 280–9.

Negueruela, I. (1995) 'Seventh-century BC Phoenician vessel discovered at Playa de la Isla, Mazarron, Spain', *International Journal of Nautical Archaeology* 24 (3): 189–97.

Negueruela, I. (1996) 'Excavaciones arqueológicas subacuaticás realizadas por el Centro Nacional de Investigaciones Arqueológicas Submarinas en el yacimiento de la Playa de la Isla (Mazarrón): Memoria de la campaña de 1995', *Memorias de arqueologia* 10: 162–80.

Negueruela, I. (2004) 'Hacia la coprensión de la construcción naval fenicia según el barco "Mazarrón 2" de siglo VII a.C.', in V. Peña (ed.), *La navigación fenicia: Tecnología naval y dirroteros; Encuentros entre marinos, arqueólogos e historiadores*. Madrid: 227–78.

Nelson, H. H. (1943) 'The naval battle pictured at Medinet Habu', *Journal of Near Eastern Studies* 2: 40–45.

Newberry, P. E. (1893) *Beni Hasan*. London.

Nibbi, A. (1991) 'Five stone anchors from Alexandria', *International Journal of Nautical Archaeology* 20: 185–94.

Olaberria, J.-P. (2013) *Hull-shape in Antiquity: how do archaeological ship remains enhance our understanding of hull-shape design in Antiquity?* PhD thesis, University of Southampton.

Olaberria, J.-P. (2014) 'The conception of hull shape by shell-builders in the Ancient Mediterranean', *International Journal of Nautical Archaeology* 43 (2): 351–68.

Pâris, E. (1841) *Essai sur la construction navale des peuples extra-européens*. Paris.

Partridge, R. (1996) *Transport in Ancient Egypt*. London.

Peachey, C. (1990) 'Model Building in Nautical Archaeology: The Kinneret Boat', *The Biblical Archaeologist* 53 (1): 46–53.

Pennington, B. and Thomas, I. R. (in preparation). 'Palaeo-landscape reconstruction at Naukratis and the Canopic branch of the Nile'.

Penttinen A., Johlman J., and Savini, E. (2005) 'The Kalaureia excavation project the 2003 season', *Opuscula Atheniensia* 30: 127–215.

Perdu, O. (2004) 'Les "blocs de Piânkhi' après un siècle de discussions', in D. Devauchelle (ed.) *La XXVIe dynastie: continuités et ruptures. Actes du Colloque international organisé les 26 et 27 novembre 2004 à l'Université Charles-de-Gaulle-Lille 3. Promenade saïte avec Jean Yoyotte*. Lille: 225–40.

Perdu, O. (2011). 'Les 'blocs de Piânkhi' après un siècle de discussions', in D. Devauchelle (ed.) *La XXVIe dynastie: continuités et ruptures. Actes du Colloque international organisé les 26 et 27 novembre 2004 à l'Université Charles-de-Gaulle-Lille 3. Promenade saïte avec Jean Yoyotte*. Lille: 225-240.

Perdu, O. (2012) 'Le prétendu "an V" mentionné sur les "blocs de Piânkhi" ', Revue d'Égyptologie 61: 151–7.

Pomey, P. (2004) 'Principles and methods of construction in ancient naval architecture', in F. Hocker (ed.), *The philosophy of shipbuilding. Conceptual approaches to the study of wooden ships*. Texas: 25–37.

Pomey, P. and Rieth, E. (2005) *L'archéologie navale*. Paris.

Pomey, P. (2012a) 'Ship remains at Ayn Soukhna', in P. Tallet (ed.), *The Red Sea in pharaonic times: Recent discoveries along the Red Sea coast. Proceedings of the colloquium held in Cairo/Ayn Sukhna 11–12 January 2009*. Cairo: 35–53.

Pomey, P. (2012b) 'The pharaonic ship remains of Ayn Sukhna', in N. Gunsenin (ed.), *Between Continents: proceedings of the twelfth symposium on boat and ship archaeology, Istanbul 2009*. Oxford: 7–15.

Pomey, P. (2012c) 'Les graffiti navals de la zone minière du Sud-Sinaï (Rod al-Air, Gebel al-Hazbar, Ouadi Shella/Ouadi Boudra', in P. Tallet (ed.), *La zone minière pharaonique du Sud-Sinaï I. Catalogue complémentaire des inscriptions du Sinaï. Le Caire: 279–96.*

Pomey, P. (2015a) 'Navires et construction navale en Égypte ancienne', in P. Tallet and B. Argémi (eds), *Entre Nil et mers. La navigation en Égypte ancienne (Nehet 3)*: 1-29.

Pomey, P. (2015b) 'La batellerie nilotique gréco-romaine d'après la mosaïque de Palestrina', in P. Pomey (ed.), *La batellerie Egyptienne. Archéologie, histoire, ethnographie*. Paris: 151–73.

Porten, B. and Yardeni, A. (1993) *Textbook of Aramaic documents from Ancient Egypt*. Jerusalem.

Pomey, P., Kahanov, Y., and Rieth, E. (2012) 'Transition from shell to skeleton in Ancient Mediterranean ship-construction: analysis, problems, and future research', *International Journal of Nautical Archaeology* 41 (2): 235–314.

Pomey, P., Kahanov, Y., and Rieth, E. (2013) 'On the transition from shell to skeleton', *International Journal of Nautical Archaeology* 42 (2): 434–7.

Porten, B. (ed.) (1996) *The Elephantine papyri in English, three millennia of cross-cultural continuity and change*. Leiden.

Powell, E. A. (2013). 'Oldest Egyptian Funerary Boat', *Archaeology* 66 (1): 31.

Pulak, C. (1987). *A Late Bronze Age Shipwreck at Ulu Burun: Preliminary Analysis (1984–1985 Excavation Campaigns)*. PhD thesis, Texas A&M.

Pulak, C. (1988) 'The Uluburun Shipwreck: an overview', *International Journal of Nautical Archaeology* 27 (3): 188–224.

Pulak, C. (2002) 'The Uluburun hull remains', in H. Tzalas (ed.), *TROPIS VII—7th International*

Symposium on Ship Construction in Antiquity. Athens: 615–36.

PULAK, C. (2008) 'The Uluburun shipwreck and Late Bronze Age trade', in J. Aruz (ed.), *Beyond Babylon. Art, trade and Diplomacy in the Second Millennium* BC. New Haven: 289–385.

PULAK, C. and INGRAM, R. (2013) 'The Shipwrecks of Yenikapı and Their Contribution to the Study of Ship Construction', in Z. Kızıltan (ed.), *Stories from the Hidden Harbor: Shipwrecks of Yenikapı*. Istanbul: 23–34.

QUIBELL, J. E. (1908) *Excavations at Saqqara (1906–1907)*. Cairo.

QUINN, R. (2006) 'The role of scour in shipwreck site formation processes and the preservation of wreck-associated scour signatures in the sedimentary record—evidence from seabed and sub-surface data', *Journal of Archaeological Science* 33 (10): 1419–32.

RAWLINSON, G. (1880) *Herodotus 'History'*. London.

RAY, J. D. (1988) 'Egypt: 525–404 B.C.', in J. Boardman (ed.), *The Cambridge Ancient History*. Cambridge: 254–86.

RIETH, E. (2008) 'Géométrie des formes de carène et construction "sur membrure première" (Ve-XIIe siècles). Une autre approche de l'histoire de l'architecture navale méditerranéenne au Moyen Age?', *Archaeologia Maritima Mediterranea* 5: 45–68.

RICHARDS, N. (2008) *Ships' Graveyards—Abandoned watercraft and Archaeological Site Formation Processes*. Gainesville.

RICHARDS, N. and Seel, S. (2013) *The Archaeology of Watercraft Abandonment*. Springer.

RIVAL, M. (1991) *La charpenterie navale romaine*. Paris.

ROBINSON, D. (2015) 'Ship 43 and the formation of the ship graveyard in the central basin at Thonis-Heracleion', in D. Robinson and F. Goddio (eds), *Thonis-Heracleion in Context*, Oxford Centre for Maritime Archaeology Monograph 8. Oxford: 211–28.

ROBINSON, D. (2018) 'The Depositinoal Contexts of Ships from Thonis-Heracleion, Egypt', *International Journal of Nautical Archaeology* 2018: 325–36.

ROBINSON, D. and GODDIO, F. (2015) 'Introduction: Thonis-Heracleion and the 'small world' of the northwestern Delta' in D. Robinson and F. Goddio (eds), Oxford Centre for Maritime Archaeology Monograph 8. Oxford: 1–12.

ROBINSON, D., GODDIO, F. and FABRE D. (2017) 'Environment and Agency in the formation of the eastern ship graveyard in the Central Basin at Thonis-Heracleion', in Caporaso, A. (ed.), *Formation processes of maritime archaeological landscapes*. Springer: 113–40.

ROGERS, E. M. (1996) *An analysis of tomb reliefs depicting boat construction from the Old Kingdom period in Egypt*. PhD thesis, Texas A&M.

ROSTOVTZEFF, M. (1940) 'Πλοια θαλασσια on the Nile', *Etudes dédiés à la mémoire d'A. M. Andréadès*. Athens: 367–76.

ROTROFF, S. I. (2011) 'Mended in Antiquity: repairs to ceramics at the Athenian Agora', in M. Lawall and J. Lund (eds), *Pottery in the Archaeological Record: Greece and Beyond. Acts of the international colloquium held at the Danish and Canadian Institutions at Athens, June 20–22 2008*. Aarhus: 117–34.

ROUGÉ, J. (1975) *La marine dans l'Antiquité*. Vendome.

ROUGÉ, J. (1987) 'La momie contenait-elle les fragments d'une voile?', *Nouvelles archives du Museum d'Histoire Naturelle de Lyon*, Fascicule 25: 91–6.

SANDERS, D. (2010) 'Knowing the ropes: The need to record ropes and rigging on wreck-sites and some techniques for doing so', *International Journal of Nautical Archaeology* 39 (1): 2–27.

SANTAMARIA, C. (1995) *L'épave Dramont "E" à Saint-Raphaël (Ve siècle ap. J.-C.)*. Paris.

SAVE-SODERBERGH, T. (1946) *The Navy of the 18th Egyptian Dynasty*. Uppsala.

SINGER-AVITZ, L. (1989) 'Local pottery of the Persian period (Strata XI-VI)', in Z. Herzog, G. Rapp Jr., and O. Negbi (eds), *Excavations at Tel Michal, Israel*. Tel Aviv: 115–44.

SLEESWYK, A. W. (1982) 'The rudder of the Blussus ship', *International Journal of Nautical Archaeology* 11: 153–4.

SLEESWYK, A. W. (1987) 'The obelisk lighter of Queen Hatshepsut', *TROPIS II. 2nd International Symposium on Ship Construction in Antiquity. Proceedings*. Delphi: 305–14.

SLEESWYK, A. W. (1996) 'Ramming trim of Ships', in H. Tzalas (ed.), *TROPIS IV. 4th International Symposium on Ship Construction in Antiquity. Proceedings*. Athens: 429–49.

SLIWA, J. (1975) *Studies in Ancient Egyptian handicraft: Woodworking*. Warsaw.

SÖLVER, C. V. (1943) *Obelisk-skibe*. Copenhagen.

SOMAGLINO, C. (2015) 'La navigation sur le Nil. Quelques réflexions autour de l'ouvrage de J. P. Cooper, The Medieval Nile. Route, Navigation, and Landscape in Islamic Egypt, Le Caire – New York, 2014', *Nehet* 3: 123–61.

SPARKES B. S. and TALCOTT, L. (1970) *The Athenian Agora: results of excavations*. Vol.XII/2. *Black and plain pottery of the 6th, 5th and 4th centuries* B.C, The American School of Classical Studies in Athens. Princeton.

SPIEGELBERG, W. (1921) *Koptisches Handwörterbuch*. Heidelberg.

STANLEY, J.-D. (2007) *Geoarchaeology. Underwater archaeology in the Canopic region in Egypt*, Oxford Centre for Maritime Archaeology Monograph 2. Oxford.

STEFFY, J. R. (1985) 'The Kyrenia Ship: An Interim Report on its Hull Construction', *American Journal of Archaeology* 89 (1): 75–101.

STEFFY, J. R. (1990) 'The Boat: A Preliminary Study of its Construction', in S. Wachsmann (ed.), *The*

Excavations of an Ancient Boat in the Sea of Galilee (Lake Kinneret). Jerusalem.

Stock, J. T. and O'Neill, M. C. (2011) 'Body size, skeletal biomechanics, mobility and habitual activity from the Late Paleolithic to the mid-Dynastic Nile Valley', in R. Pinhasi (ed.), *Human bioarchaeology of the transition to agriculture*. Chichester: 347–67.

Tallet, P. (2012) 'Ayn Sukhna and Wadi el-Jarf: Two newly discovered pharaonic harbours on the Suez Gulf', *The British Museum Studies in Ancient Egypt and Sudan* 18: 147–68.

Tallet, P. (2013a) 'The Wadi el-Jarf site: A harbor of Khufu on the Red Sea', *Journal of Ancient Egyptian interconnections* 5 (1): 76–84.

Tallet, P. (2013b) 'Deux notes sur les expéditions au pays de Pount', *Revue d'Égyptologie* 64: 189–210.

Tallet, P. (2015) 'Les "ports intermittents" de la mer Rouge à l'époque pharaonique: caractéristiques et chronologie' in P. Tallet and B. Argémi (eds), *Entre Nil et mers. La navigation en Égypte ancienne. Nehet* 3: 31–72.

Tallet, P. and Mahfouz, E. (2012) *The Red Sea in pharaonic times: Recent discoveries along the Red Sea coast. Proceedings of the colloquium held in Cairo/Ayn Sukhna 11–12 January 2009*. Cairo.

Tallet, P. and Marouard, G. (2014) 'The harbor of Khufu on the Red Sea Coast at Wadi al-Jarf, Egypt', *Near Eastern Archaeology* 77 (1): 4–14.

Tallet, P., Marouard, G. and Laisney, D. (forthcoming). *Un port de la IVe dynastie au ouadi el-Jarf (mer Rouge)*. Cairo.

Thomas, I. R. (2015) 'Naukratis, "Mistress of ships" in context', in D. Robinson and F. Goddio (eds), *Thonis-Heracleion in context*, Oxford Centre for Maritime Archaeology Monograph 8. Oxford: 247–66.

Thompson, D. and Garnsey, P. (1983) 'Nile Grain Transports under the Ptolemies', *Trade in the Ancient Economy*. London: 64–75.

Tzalas, H. (1999) 'Were the pyramidal stone-weights of Zea used as anchors?' in H. Tzalas (ed.), *TROPIS V—5th International Symposium on Ship Construction in Antiquity*. Athens: 429–55.

Vaag L. E., Norskov V. and Lund, J. (2002) *The Maussolleion at Halikarnassos 7, The Pottery. Ceramic Material and others Finds from Selected Contexts*. Aarhus.

Van't Dack, E. (1978) 'L'apport égyptien à l'armée navale lagide', in H. Hauben (ed.), *Das Ptolemäische Ägypten, Akten des Interinationalen Symposiums 27–29 September 1976, Berlin*. Mayence.

Veldmeijer, A. J. (2009) 'Cordage production', in W. Wendrich (ed.), *UCLA encyclopedia of Egyptology*. Los Angeles.

Villing, A. (2015) 'Egyptian-Greek exchange in the Late Period: the view from Nokradj-Naukratis', in D. Robinson and F. Goddio (eds), *Thonis-Heracleion in context*, Oxford Centre for Maritime Archaeology Monograph 8. Oxford: 229–46.

Vinson, S. (1993) 'The Earliest Representations of Brailed Sails', *Journal of the American Research Center in Egypt* 30: 133–50.

Vinson, S. (1994) *Egyptian Boats and Ships*. Buckinghamshire.

Vinson, S. (1996) 'PAKTOUN and PAKTOSIS as Ship-Construction terminology in Herodotus, Pollux and Documentary Papyri', *Zeitschrift für Papyrologie und Epigraphik* 113: 202.

Vinson, S. (1997) 'On *hry.t* "Bulwark", in P.Anastasi IV, 7/9-8/7', *Zeitschrift für Ägyptische Sprache* 124 (2): 156–62.

Vinson, S. (1998) 'Remarks on Herodotus' Description of Egyptian Boat Construction (II, 96)', *Studien zur Altägyptischen Kultur* 26: 251–60.

Wachsmann, S. (1981) 'The ships of the Sea Peoples', *International Journal of Nautical Archaeology* 10 (3): 187–220.

Wachsmann, S. (1987) 'The Kinneret boat: the discovery and excavation', in H. Tzalas (ed.), *TROPIS* II. *2nd International Symposium on Ship Construction in Antiquity. Proceedings*. Delphi: 371–84.

Wachsmann, S. (1989) *Seagoing ships and seamanship in the Late Bronze age Levant*. PhD thesis, Hebrew University.

Wachsmann, S. (ed.) (1990) *The excavations of an ancient boat from the Sea of Galilee (Lake Kinneret)*. Jerusalem.

Wachsmann, S. (1995) 'The Kinneret boat: the excavation report', in H. Tzalas (ed.), *TROPIS* III. *3nd International Symposium on Ship Construction in Antiquity. Proceedings*. Athens: 471–85.

Wachsmann, S. (2009) *Seagoing ships and Seamanship in the Bronze Age Levant*. College Station.

Wachsmann, S. (2010) 'Ahhotep's silver ship model: the Minoan context', *Journal of Ancient Egyptian interconnections* 2 (3): 31–41.

Wachsmann, S. (2013) *The Gurob ship-cart model and its Mediterranean context*. College Station.

Wachsmann, S., Raveh, K. and Cohen, O. (1987) 'The Kinneret Boat Project. Part I. The excavation and the conservation of the Kinneret boat', *International Journal of Nautical Archaeology* 16 (3): 233–45.

van der Wilt, E. M. (2013) *The Place of Lead in an Egyptian Port City in the Late Period*. DPhil dissertation, Oxford.

Ward, C. (2000) *Sacred and secular: Ancient Egyptian ships and boats*. Boston.

Ward, C. (2003) 'Sewn planked boats from Early Dynastic Abydos, Egypt', in C. Beltrame (ed.), *Boats, ships and shipyards. Proceedings of the Ninth International Symposium on Boat and Ship Archaeology, Venice 2000*. Oxford: 19–23.

Ward, C. (2004) 'Boatbuilding in Ancient Egypt', in

F. Hocker (ed.), *The philosophy of shipbuilding. Conceptual approaches to the study of wooden ships*. Texas: 13–25.

Ward, C. (2007) 'Ship timbers: description and preliminary analysis', in K. A. Bard (ed.), *Harbour of the pharaohs to the Land of Punt. Archaeological investigations at Mersa/Wadi Gawasis, Egypt, 2001-2005*. Naples: 135–50.

Ward, C. (2009) 'Evidence for Egyptian seafaring', in R. Bockius (ed.), *Between the seas: Transfer and exchange in nautical technology. Proceedings of the Eleventh international symposium on boat and ship archaeology. Mainz, 2006*. Mainz: 9–16.

Ward, C. and Zazzaro, C. (2010) 'Evidence for Pharaonic seagoing ships at Mersa/Wadi Gawasis, Egypt', *International Journal of Nautical Archaeology* 39 (1): 27–44.

Ward, C. and Couser, P. (2012) 'Reconstruction and sailing performance of an ancient Egyptian ship', in N. Gunsenin (ed.), *Between Continents: proceedings of the twelfth symposium on boat and ship archaeology, Istanbul 2009*. Oxford: 287–93.

Wegner, J. (2016) 'A Royal Boat Burial and Watercraft Tableau of Egypt's 12th Dynasty (*c*. 1850 BCE) at South Abydos', *International Journal of Nautical Archaeology* 46 (1): 5–30.

Wehausen, J. V., Mansour, A. and Ximenes, M. C. (1988) 'The colossi of Memnon and Egyptian barges', *International Journal of Nautical Archaeology* 17 (4): 295–310.

Weisman, A. D. (1899) Греческо-русский словарь, Москва.

Werner, E. K. (1986) 'Montu and the "Falcon Ships" of the Eighteenth Dynasty', *Journal of the American Research Center in Egypt* 23: 107–23.

Westerdahl, C. (1992) 'The Maritime cultural landscape', *International Journal of Nautical Archaeology* 21 (1): 5–14.

Whitbread I. K. (1995). *Greek Transport Amphorae. A Petrological and Archaeological Study*, Fitch Laboratory Occasional Paper, 4. Athens.

Wild, F. C. and Wild, J. P. (2001) 'Sails from the Roman port at Berenike, Egypt', *International Journal of Nautical Archaeology* 30: 211–20.

Yoyotte, J. (1958) 'Notes de toponymie égyptienne', *Mitteilungen des deutschen archäologischen instituts, Röm. Abt.* 16: 414–30.

Yoyotte, J. (1994) 'Les contacts entre Egyptiens et Grecs (VIeI–IIe s. av. J.-C.): Naucratis, ville égyptienne', *Annuaire du Collège de France. Résumé des cours et travaux 1992–1993, 1993–1994*: 679–92.

Yoyotte, J. (2001) 'Le second affichage du décret de l'an 2 de Nekhtnebef et la découverte de Thônis-Héracleion', *Egypte, Afrique & Orient* 24: 24–34.

Yoyotte, J. (2013) *Histoire, géographie et religion de l'Egypte ancienne: Opera Selecta*. Leuven.

Zazzaro, C. (2007) 'Ship blades, anchors and pierced stones', in K. A. Bard (ed.), *Harbour of the pharaohs to the Land of Punt*. Naples: 150–60.

Zazzaro, C. (2009) 'Nautical evidence from the pharaonic site of Marsa/Wadi Gawasis. Report on two parts of a steering oar/rudder', in R. Bockius (ed.), *Between the seas. Transfer and exchange in nautical technology. Proceedings of the 11th international symposium on boat and ship archaeology, Mainz 2006*. Oxford: 3–8.

Zazzaro, C. (2011) 'Les ancres de Mersa Gawasis', *Egypte, Afrique & Orient* 64 (2): 13–20.

Zazzaro, C. and Abd El-Maguid, M. (2012) 'Ancient Egyptian stone anchors from Mersa Gawasis', in P. Tallet (ed.), *The Red Sea in pharaonic times: Recent discoveries along the Red Sea coast. Proceedings of the colloquium held in Cairo/Ayn Sukhna 11–12 january 2009*. Cairo: 87–105.

Index

The sunken portion of the ancient canopic region shown on a satellite picture of the west delta of the Nile.